Education and training of users of scientific
and technical information:
UNISIST guide for teachers

A. J. Evans
R. G. Rhodes
S. Keenan
Loughborough
University
of Technology
United Kingdom

Education and training of users of scientific and technical information

UNISIST guide for teachers

unesco

The designations employed and the presentation of material in this work do not imply the expression of any opinion whatsoever on the part of the Unesco Secretariat concerning the legal status of any country or territory, or of its authorities, or concerning the delimitations of the frontiers of any country or territory.

Published in 1977 by the United Nations
Educational, Scientific and Cultural Organization
7 Place de Fontenoy, 75700 Paris
Printed by Offset Aubin, Poitiers (P 7281)

ISBN 92-3-101452-8
French edition: 92-3-201452-1
Spanish edition: 92-3-301452-5

Preface

Unesco's concern in improving the international transfer and exchange
of scientific and technological information among its Member States
led to the launching of UNISIST in 1972 - an intergovernmental programme
for co-operation in the field of scientific and technological information.

Basic to the philosophy of the UNISIST programme is the belief that the
final goal of every single library, documentation centre, information
service, national information system or regional network is to serve the
users of information. Users of information are an integral part of and
the final link in the information transfer chain; they are the 'raison
d'être' of every investment made for bettering the storage, processing
and retrieval of information.

In every country, from the least developed to the most advanced, a
certain amount of scientific and technological information - locally
produced or of international origin - is processed and stored in some
fashion for the benefit of users. Unless these users know how to find
relevant information available to them, the information 'machinery' falls
short of its main goal.

Users may be trained at various levels and in different ways. Perhaps the
most important way is through formal courses or workshops for developing
information-retrieval skills in users. This Guide is intended for teachers
concerned with developing and conducting such courses.

The preparation of this Guide was entrusted to Professor A. Evans of
Loughborough University, assisted by Mr. R.G. Rhodes and Miss S. Keenan
of the same institution. They have benefited from the co-operation of an
International Advisory Board specially set up by Unesco for this purpose.
The authors of this publication and all those who have assisted in its
preparation are to be commended for their efforts and dedication.

This is only the first edition of what must be regarded as an initial
effort. Comments from any source which will help Unesco to present in
later editions a more complete, coherent and practical guide will be
welcomed, and should be addressed to the Division of Scientific and
Technological Documentation and Information.

Contents

Table of Contents

Table of Contents

Foreword

The intended readers of this Guide are the teachers who develop information-retrieval skills in users of scientific and technical information. These teachers may be:

1. Working in their own institution, at which students study scientific or technical subjects at graduate or undergraduate level;

2. Organizing at their own institution short courses for working scientists or technologists who attend the institution for the duration of the courses;

3. Visiting an institution to teach at special courses to that institution's undergraduate or graduate students;

4. Visiting an institution to organize a short course for working scientists and technologists who attend the institution for the duration of the course.

The Guide is essentially based on the authors' experience of running undergraduate and postgraduate courses at Loughborough University of Technology over the last ten years. It has also been tested away from the teachers' normal environment. Before it reached its final version, workshops were held in Yugoslavia (essentially type 2 as listed above), Indonesia (mixture of 2 and 4), India (both 3 and 4) and South Korea (type 4). The experience gained has been incorporated in the Guide, where applicable; it clearly emphasises the need for a totally flexible approach from initial planning to the final day of a course.

The Guide is designed to cover all four types of course and the user (i.e. course teacher/organiser) should first identify which type of course is to be planned. To use it successfully, he should be as familiar as possible with the following:

(a) the institution in which the course is to be held, and its library and information collection;

(b) the target audience for the course, level, subject area, etc;

(c) information-retrieval techniques in general and specialized reference tools.

The Guide is not intended to be a "teach yourself information retrieval" handbook. It is specifically written for people responsible for organizing and running training courses of the kinds listed above. It should be used in conjunction with the recently issued UNISIST guidelines.[1] Some of the material can be used as handouts as it stands or in a modified form. The authors realize that it is impossible to cover all aspects of, and approaches to, user training exhaustively. They hope that course organizers will find in the Guide the basic structure and planning needed for course organization, which can be expanded according to the teachers' own knowledge and experience and the particular circumstances under which a specific course may be developed.

They will certainly find on occasions insufficient detail to meet particular requirements on a given course (for example, in the discussion of mechanized information-retrieval systems in Section 7.4.3). A list of suggested items for further reading has therefore been included as Appendix 5. There has been no attempt to produce a bibliography on user education, since bibliographies are available from many other sources.[2]

The authors wish to thank most sincerely G.E. Hamilton, C.M. Lincoln, K. Phillips, R.A. Wall and E.A. Wilkinson for their contribution and assistance in the preparation of this Guide and J. Anderson and C.M. King for typing the manuscript.

Loughborough
May, 1976

A.J. Evans
R.G. Rhodes
S. Keenan

---

[1] "Guidelines for the Organization of Training Courses, Workshops and Seminars in Scientific and Technical Information and Documentation." Prepared by Pauline Atherton. Paris, Unesco. 1975. 88p. (Doc.SC/75/WS/29)

[2] Such as John Lubans: "Educating the Library User", New York, Bowker. 1974. 435p

Acknowledgements

The authors wish to thank the following people for their critical and stimulating comments on the draft : P. Atherton (USA), K.P. Barr (United Kingdom), C. Bourne (USA), A. Debons (USA), Doo-hong Kim (Korea), H.E. Gomes (Brazil), V.I. Gorkova (USSR), C. Hanson (United Kingdom), S. Herner (USA), P. Judge (Australia), A-H.K. Jumba-Masagazi (Uganda), C. Keren (Israel), K. Klintøe (Denmark), P. Luwarsih (Indonesia), R.T. de Mautort (UNIDO), M. Menou (France), J. Meyriat (France), E. Molino (Mexico), S. Parthasarathy (India), J. Quevedo P. (Mexico), R.S. Rajagopalan (India), K. Samuelson (Sweden), W.L. Saunders (United Kingdom), M. Slajpah (Yugoslavia), J.I. Snyman (Rep. of South Africa), L. Vilentchuk (Israel), P. Wasserman (USA), G. Wersig (Fed. Rep. of Germany), R. Wettling (France), D.E.K. Wijasuriya (Malaysia), F.K. Willenbrock (USA).

1. Introduction

Man is a strange creature, and many of his motives and actions are illogical and emotional. Some resources available in relative abundance have been overlooked or neglected. Time is now recognized as a resource - an even greater resource than money - and it is still being wasted. Air, water, space and other components of the environment are used in industrial processes but are not fully costed. These components are apparently discharged or released after serving their useful purpose, but often in an inferior condition. Pollution is now widely acknowledged as a degradation of the environment.

In a similar way information is a resource which is abused, wasted, neglected and degraded. The acquisition of information is often not costed and is still thought by many people to be free. Time, environment, materials, manpower, finance and information are resources and it is important internationally, nationally and individually to understand their nature, use and benefits.

Generally, information - like air - is available throughout the world; national boundaries have only a relatively small effect on its availability. Thus information can be regarded as an international resource.

The teacher of information retrieval should understand the overall pattern of human communication and information flow. In general the appropriate procedure is to question existing concepts and to stimulate rethinking of basic approaches. Each individual must determine for himself his own concepts.

For the purpose of this Guide, information is taken as a sensible concept, statement, or idea, or an association of concepts, statements or ideas. When information is stored in the mind it constitutes knowledge, particularly when relationships are established between items of information. When the significance of the relationships is fully known, it constitutes understanding. The distinctions between information, knowledge and understanding are vague, although the philosopher might not agree.

This use of the term 'information' accepts that information is synthesized from items of knowledge. The mental ability to grasp relationships between items of information is usually referred to as intelligence.

Although information itself is intangible, it can be collected, recorded and stored - mainly the written word and preferably printed. Such documentary information or literature can be made available as a commodity. Although the other forms of recorded information, such as magnetic-tape and films, add complexity to the situation, they do have a role to play and this is likely to increase.

Distribution of documentary information is carried out by a combination of publishers, booksellers and libraries. Those who purchase, make reference to and compile publications also take part in the distribution of recorded information.

Libraries are the stores of information; the users are frequently the same community as the producers of information. If the content or quality of the available literature is found unsatisfactory, that community must take a substantial portion of the blame.

For the effective use of the available literature basic skills need to be developed, for which librarians and teachers must accept some responsibility. It is the need for these skills which makes libraries a somewhat under-utilized group of resources. However, it should be emphasized that not all information is available as literature, since some is regarded as private by individuals, organizations or nations. The amount of it is small relative to the amount of literature; but it contains two notable categories - financial and management information.

An individual with an information need goes through the following processes. He searches his own memory for relevant earlier experience, questions colleagues and specialists, and searches the literature. Most people are not conscious of the extent and range of the world's literature. Many find it difficult to grasp the extent because of its sheer magnitude. Few recognize that some published material is of little value, probably not worth the effort of writing and publication. The presence of this lower-quality material makes it essential to identify and select the useful material methodically.

For the scientist and technologist learning the basic skills of information retrieval, it is probably sufficient to describe the value and the various sources of information. Ideally, it would be accompanied by instruction on such topics as report writing, research methods and general communication.

1.1 AIM OF THE GUIDE

This Guide has been written for people who have the task of teaching information retrieval to users of scientific and technical information. These users may be graduate or undergraduate students in particular scientific and technical subjects or they may be working scientists and technologists seeking to upgrade or update their skills in information retrieval. The Guide is intended to have particular application in developing countries.

Teaching information retrieval to these users can be difficult and it is easy to misjudge one or more of the factors. The teacher is likely to need expertize in library and information science, acquaintance with subject areas, teaching skills and knowledge of local circumstances. Individual teachers may be strong in some of these requirements but weak in others. For example, a librarian may gradually slip into an approach more suitable for teaching student librarians. On the other hand, a subject teacher may be too orientated towards the subject specialization.

If one assumes that the teacher has the necessary knowledge of library and information science and appropriate subject expertise, the first problem is to motivate the target audience for the training course. Many people do not appreciate the significance of information retrieval or even recognize it as relevant to their studies or profession. Further difficulties may be created when groups of mixed backgrounds and differing degrees of comprehension are involved.

The second problem is how to describe logical and systematic methods of retrieving information. It arises because the literature cannot be handled in a totally logical and systematic manner. The solution is sometimes a compromise and sometimes complex. The Guide aims to provide guidelines for dealing adequately with the compromises and complexities while seeking information in a logical and systematic manner.

No single approach or course content is appropriate for all groups. On the contrary, each need deserves separate consideration. Local circumstances can influence both approach and content. Factors such as jobs or fields of study, inter-library co-operation, group size, environment, related work and demands on time need to be considered. Many of these factors can be anticipated and incorporated into the planning. Sometimes unexpected events make it necessary to discard planned material and to substitute more appropriate material during the actual presentation.

This Guide provides guidelines which, it is hoped, can help overcome the problems described above.

1.1.1 Definitions of terms

Many of the terms used in information retrieval and its related fields have a number of interpretations. Their meanings as used in this Guide are as follows:

INFORMATION - a sensible statement, opinion, fact, concept or idea, or an association of statements, opinions or ideas. It is closely associated with knowledge in that once information has been assimilated, correlated and understood it becomes knowledge.

INFORMATION GATHERING - the process of collecting and receiving information by any available means including recall from memory, deduction, communication with other people (including reading published material), and detection or observation by any of the physical senses assisted by experiments.

INFORMATION RETRIEVAL - that part of information gathering which is concerned with the efficient selection of items of existing knowledge and information, particularly that which is stored in a recorded form.

INFORMATION HANDLING - a term used increasingly for what has traditionally been called information retrieval. Although a near synonym for information retrieval, this topic includes part of the processing of information once it has been gathered, for example, the creation of personal files.

INFORMATION MANAGEMENT - has nearly the same meaning and application as information handling. It is naturally appropriate for the instruction of management students. In practice the term implies the inclusion of management information systems, which is itself an extensive topic and not included in the scope of this Guide.

COMMUNICATION - the process of transferring information involving a source, a transmission medium or method, and one or more receivers. Each component is significant in that the source should express the information clearly, the medium should convey the information efficiently, and the receiver should understand the information. This implies responsibilities for both the communicator and the receiver of information. Traditionally the writer's and speaker's responsibilities have been recognized but not always adequately fulfilled. The role and responsibilities of the reader have tended to be vague and he is sometimes reluctant to play an active role in the process of acquiring information. Thus passive sources of information such as publications have been undervalued and under-used. The teaching and use of information-retrieval skills can help to ensure efficient and effective communication.

COURSE - a series of lectures or instruction for a particular group of people, normally described in a syllabus or time-table and possibly involving an assessment of performance. Information retrieval or communication may be the subject of a short course of one or more days. In longer courses information retrieval can be integrated with associated topics.

TUTOR - a tutor is a teacher. The word, like 'instructor', usually implies special circumstances or duties. Thus a course tutor is concerned with directing the studies of students on a particular course.

1.2 TARGET GROUPS

In general, this Guide should be of value to any teacher concerned with teaching information retrieval to any group of users of scientific and technical information. The principal exception is a group of practising and student librarians, who will recognize that information-retrieval studies have a different role in their field.

The broad target group consists of scientists and technologists who are studying for or possess degrees or professional qualifications and who have not before received any significant instruction in information retrieval.

In practice, many groups have more specific characteristics, interests or educational levels. Thus the teacher needs to modify the basic material according to the common features. Notes for particular subject groups are given at appropriate points in this Guide, especially sections 5.2.2, 6.3.1 and 7.5.1, which deal with the subject differences of the following specific target groups:

General Science

Mathematics and Computer Science
Astronomy
Physics
Chemistry
Geology
Life Sciences

Engineering and Technology in General

Medicine
Mechanical, Civil and Production Engineering
Electrical and Electronic Engineering
Aeronautical and Astronautical Engineering
Automobile Engineering
Agriculture, Horticulture, Forestry and Land Management
Management
Chemical Engineering
Food Science and Technology
Materials, Science and Technology

Social Sciences, Humanities and Arts in General
Social Sciences
Economics
Humanities
Arts

Sections 5.2.3, 6.3.2 and 7.5.2 deal, in a similar manner, with differences in educational level:

Undergraduate studies,

Postgraduate studies,

In-service, further or continuing education.

1.3 STRUCTURE OF THE GUIDE

Once the reader is familiar with the aims, scope and structure of the Guide, he can consult it as appropriate. Following this introductory chapter, the Guide is divided into two parts:

Part I - Chapters 2 to 4 : preliminaries and preparation,
Part II - Chapters 5 to 9 : presentation and content.

In these chapters attention is often focussed on one of the three principal target groups:

(a) undergraduate students,
(b) postgraduate students,
(c) practising scientists and technologists.

In some chapters, however, particularly those concerned with preliminaries and preparation, undergraduate and postgraduate student needs are dealt with together and only practitioners separately.

Some sections are intended for particular circumstances and usually exist as supplementary notes towards the ends of chapters, particularly those concerned with presentation and content.

Normally the teacher is expected to have a familiarity with some of the local circumstances, including a knowledge of the available stock and services in the library or information centre. For the visiting teacher this familiarity may be lacking and can be a serious obstacle. Consequently section 3.4 has been devoted to this problem.

Certain parts of the text can be used as actual lesson notes or content. Such parts are indented, as in the definitions section 1.1.1 which occurred earlier in this section.

Part I Preliminaries and preparation

2. Benefits from educating the user

In educating users of scientific and technical information one has first to establish the benefits of using available information as effectively as possible and then to consider the objectives of the training course, the motivation of those taking the course and the practical arrangements that have to be made.

The following figure shows how these aspects are covered in Chapters 2 - 4:

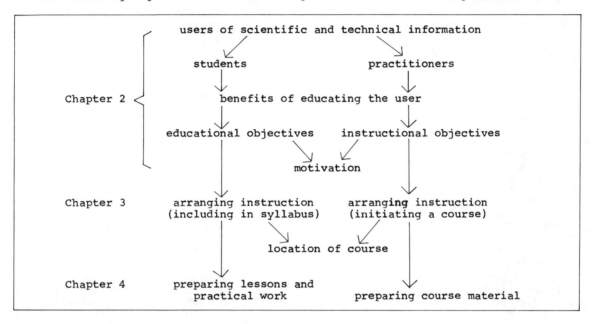

Figure 2.0 Connections between Chapters 2, 3 and 4

In establishing the benefits of using information one needs to draw on one's own knowledge and on published studies, and to supplement these by particular case studies, especially if they are relevant to potential study groups. Much of the published material is on general principles or major cases. The more routine use of information, though at least as important, is not given the same attention. Discussions with practitioners help to identify and describe routine use, though practitioners may often overlook routine steps in acquiring and using information. It should then be possible to summarize the benefits of using information in a brief statement on which the educational (or instructional) objectives can be based.

Teachers who have practised as scientists or technologists will appear to be at an advantage in studying the main users of scientific and technical information, because they have gained insight into the problems and actions of their colleagues. Such insight is, however, insufficient, because each teacher needs to appraise the total role of information in science and technology, including the potential benefits in a developing situation.

Scientists and technologists can be said to provide society with information translated into knowledge by skill and also with material products. In return, they seek some financial reward, a degree of satisfaction and the esteem of their colleagues.

There is little relationship between scientific effort, contribution, and reward. An intellectually demanding investigation may produce an elegant solution to a problem, but the contribution and reward are likely to be small if there is unwitting duplication of a documented earlier study. On the other hand, a potential contribution may go unrealized and unrewarded by stagnating in an unreadable or unretrievable form. The greatest financial reward does not always go to the discoverer of an exciting breakthrough, but more often to the exploiter of a discovery which can have application in most households.

It seems desirable that any instruction in information retrieval should improve the beneficial contribution of scientists and engineers to society. This improvement should be sufficient to justify the time and energy spent on developing retrieval skills.

The ways in which the contributions of scientists and technologists are affected by information-retrieval skills are widely accepted. They include avoidance of unnecessary duplication of research or measurement; building on existing knowledge (sometimes known as "hitch-hiking" or "standing on the shoulders of giants"); avoiding fruitless lines of research; and applying the techniques and developments of one field to another ("cross-fertilisation").

It is worth the teacher's while to study actual cases and form an impression of the key factors. The studies can be grouped in four categories:

1. a) The use or non-use of information in successful scientific or technological developments.
   b) The use or non-use of information in failed scientific or technological developments.

2. a) The correlation between use of libraries and information centres and personal scientific contribution.
   b) The correlation between non-use of libraries and information centres and personal scientific contribution.

1(a) Studying successful scientific and technological developments.
Published studies can be of interest and value, but they have limitations. One is that the objectivity and selective approach of the writer can result in insufficient detail on the use of information. Another is that the development of a now familiar item such as the ball-point pen creates wide interest whereas an obscure but vital piece of highly technical equipment will have limited interest.

The better studies are objective, and they may refer to other useful studies. Also information services sometimes record the fact that information from the services has reduced costs and influenced product development. Discussions with practising scientists and engineers will produce examples, as will regular scanning of general scientific and technological journals.

1(b) Studying failed scientific and technological developments.
Some studies include comparisons between successful and failed projects and these can be particularly useful. Otherwise, some published material results from enquiries into serious failures, such as a collapsed bridge. The studies are detailed and objective and give perhaps the clearest indication of the potential role of information-retrieval skills and its correlation with other factors such as communication, human failings and predictability. Most of the reports are by official bodies and are fairly easily obtainable, though subject to government practice in the country of origin.

2(a) Studying correlation between use of libraries and information centres and personal scientific contribution.
It is well known that different people use libraries and information centres to different extents. While substantial use does not guarantee success, studies have yielded interesting findings and the teacher should be familiar with some of them, particularly on the role of people who supply selected information to colleagues and on informal direct contacts. Examination of the loans file of a local library or the use pattern of an information service can be interesting.

2(b) Studying correlation between non-use of libraries and information centres and personal scientific contribution.
Use and non-use are very closely linked because, for example, one person may seek or provide information on behalf of another. However, there is a tendency to overlook people who do not use libraries and information centres and to concentrate on those who do. The reasons for non-use are as important as those for use; and both potential and actual use should be studied.

In collecting information in these areas one must remember the changing pattern of scientific and technological development and the growth of recorded literature. It should also be noted that the need to educate users has become apparent only in the past thirty years and actively pursued only during the past fifteen to twenty years.

In scientific development there is an increasing number of scientists and an increasing amount of scientific knowledge. This growth is balanced to some extent by specialization and by a better understanding of scientific principles; but both factors make communication important, through formal channels (written papers) and informal channels (meetings and discussions).

In technological development the time-lag between innovation and marketing is important, as is an early awareness of new advances. Consequently it is now more important to use several good channels of communication.

Whether the teacher normally practises as a scientist or technologist or teaches full time, there is a danger of becoming complacent about information resources. The full-time teacher may not be in touch with the latest developments in information retrieval and a working scientist or technologist may rely too much on personal contacts to satisfy his information needs. With the development of computer files of references and data that may be stored in remote locations and then used on-line by telecommunications, recorded information is no longer restricted to the printed word. Both types of source need to be fully explored. It can also be a valuable exercise to examine carefully and list the changes that have occurred over a ten-year period in such finding tools as Encyclopaedia Britannica, Chemical Abstracts or Science Citation Index.

Potential User Benefits

1. Higher standards and higher productivity result from the development of the ability to study a chosen topic independently, gain awareness of existing knowledge in a specified area, acquire detailed data and comparable factual statements and be aware of new developments of potential value. There should also be less duplication of effort or repetition of failures.
2. More efficient use is made of available literature, because users can search directly when either necessary or advantageous. In addition awareness of the possibilities and problems of information retrieval makes for better indirect use - through external information services.
3. The effective use of information indirectly advances technological development in the national interest.

Potential Library and Information Centre Benefits

1. Users who are more self-reliant enable staff to concentrate on areas in which they can have the greatest effect.
2. Greater productive use makes improvements and developments more likely for the parent institution.
3. Improved understanding of information-retrieval methods and services by users makes it easier for the institutions to acquire new and up-dated reference services.

2.1 EDUCATIONAL AND INSTRUCTIONAL OBJECTIVES

Educational and instructional objectives vary according to the beliefs of the teachers, which reflect variations in understanding of the benefits of information-retrieval skills and which influence the approach to, and content of, lessons and courses. The teacher needs to set down the educational or instructional objectives for each group according to his own beliefs and the circumstances.

His starting point is the identified benefits of educating the user. He can then consider the background and aspirations of the individuals who make up the course. It is then necessary to establish the main objectives:

1. To enable the individuals to do immediately after the instruction tasks that they could not do well, if at all, prior to instruction. (HAVING INFORMATION-RETRIEVAL SKILLS.)
2. To secure a change in behaviour or attitudes resulting from the instruction. (USING INFORMATION-RETRIEVAL SKILLS.)
3. To secure an embodiment of the skills as a fairly permanent feature of work. (SATISFACTORY RESULTS FROM INFORMATION-RETRIEVAL SKILLS.)

Much has been written in educational literature on gaining and using skills, and details have been categorized and studied, together with methods of assessing the extent to which objectives have been achieved. If assessment is made soon after instruction, the results are usually more optimistic than after the lapse of several months, during which more practical experience - both beneficial and frustrating - can greatly influence achievement. Hence the third specific objective - satisfaction.

The teacher of information retrieval should pursue the educational and instructional objectives which he believes in after studying relevant aspects of the use and value of information.

Established objectives and existing methods of educating users should be submitted to simple tests:

1. Can the existing instruction result in improved performance of users?
2. Is the existing instruction accepted and used?

The educational objectives need to be clearly in the teacher's mind in order to avoid digressions and to keep the content pertinent to the needs of the students. However, the objectives should be flexible and should vary according to understanding of the general need for information retrieval and the specific needs of different classes and levels of user.

When establishing the educational objectives for a group of users, one should ideally be able to visualize or observe the daily practices of the group, whether university or college students or practising engineers and scientists. Unfortunately, however, information needs are often recognized only when difficulties or failures are experienced. Someone visiting another country to teach information retrieval might well examine the sort of situation which occurs there as well as considering available bibliographical sources.

The teacher's aim should be to bridge the gulf which exists between the available information and the users' areas of interest - not necessarily the immediate areas of interest but possibly those anticipated over a period of time. Visits to a local manufacturer or research station can indicate to the teacher the problems which present-day undergraduates may meet in a few years' time. The visits need not be numerous or frequent, but should enable the teacher to understand the context of the students' information needs. The problems and, if possible, the solutions should be considered. If the teacher can show that the problems can be more quickly, or better, solved by using the available published sources, directly or indirectly, so much the better.

If the teacher can gain this insight into the needs, roles and attitudes of a specific group, he can fairly easily adjust the educational objectives and create sufficient motivation to achieve them. In practice it is not possible to do this for all groups and he must often resort to generalizations.

2.1.1 Developing Countries

The circumstances which exist in developing countries probably resemble those existing in any other country, in that some workers will be orientated to specific objectives and very practical in their approach, while others will be pursuing ideas for their academic value. In developing countries there is likely to be a higher emphasis on practical objectives because these will be linked to areas of development, and it is possible that local circumstances will influence specific projects. Such circumstances may be geographical - eg. in a nation of islands with a limited transport system attempting to exploit and export a local crop which deteriorates fairly rapidly.

2.1.2 Subject Differences

Subject differences will influence the enrichment of the basic educational objectives rather than the objectives themselves. Thus, scientists will be interested in the pursuit of knowledge, particularly the creation of new ideas and the discovery of new relationships. They will wish to know what has been done before and what current lines are being pursued. The working data will be the accepted values of established phenomena, for a new line of thought may be stimulated by a deviation from normal values. Technologists will apply the ideas and discoveries of scientists to satisfying the material needs of mankind. In addition, they will meet new problems and will consider the available information in finding the best solutions.

2.1.3 Differences in Target Groups

The differences are mainly of need and of ability to appreciate the significance of material. Whereas a group of practitioners may have need for early information on innovations and so can appreciate and use help in obtaining information from patents, university students may find patents of interest but difficult to use. Practitioners may find the writing of essays boring but should want to improve their productive skills. It is usually appropriate to refer to objectives as educational for student groups and instructional for practitioners.

Undergraduate Groups. The educational objectives of teaching information retrieval should be put into context of the total educational objectives of the institution and its members. This ensures co-operation, and students can more easily appreciate the significance of the skills they acquire since they are often very single-minded in seeking qualifications.

Postgraduate Groups. Postgraduates have more experience and maturity than undergraduates but their needs are at a higher level and they have an earlier need for skills associated with the practice of their profession.

Groups of Practitioners. The differences between postgraduates and practitioners are that the latter are essentially vocational in orientation and that their attitudes are based on the reasons for attending the course and on the nature of their work. Their common features include a high practical orientation, and interest in solving problems, in increasing their contribution and in status.

2.2 MOTIVATION

Motivation is closely linked to the attitude of the people taking the course and is one of the important recurring topics which affect both planning and content.

2.2.1 Motivation in Planning

The various aspects of motivation in planning can be illustrated by a simple diagram, as shown in Figure 2.2.1.

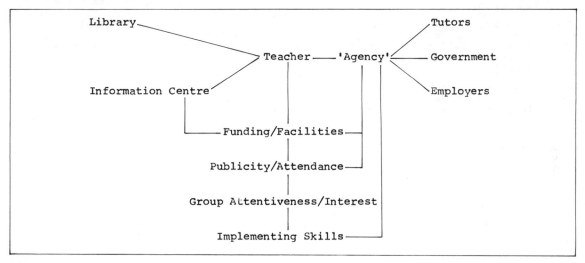

Figure 2.2.1 Aspects of Motivation in Planning

The benefits of information-retrieval skills occur when they have been used to good effect. As the diagram shows, there are several stages between the concept of instruction and the implementation of skills. At each stage motivation may or may not exist and the planning may need to be different for each circumstance.

The concept of instruction and the initial discussions may come from the teacher, the library/information centre, or an educational or governmental agency. One or more of these sources need to be sufficiently motivated to take the initiative.

2.2.2 Motivating the Teacher

Motivation needs to be strong in the teacher, whose enthusiasm and knowledge has to convince people taking the course of the significance of information retrieval in their vocation and field of study. Ideally the individual needs to be self-motivated; that is, to desire to do the teaching. A visiting teacher needs co-operation from the local library or information centre, which needs to be motivated to work for the success of the course.

2.2.3 Motivation from the Library/Information Centre

It is important for the teacher first to consider the library or information centre where the course will be held. Its facilities are essential to instruction, and if they are inadequate on important points it is better to delay instruction until adequate provision is made. Until instruction is generally accepted, there is a need to stimulate and interest readers; but if frustration results from inadequate facilities then negative motivation will occur.

Ideally the facilities should be extensive, so that the full potentialities of information retrieval be demonstrated. In practice the facilities need to be relatively few, particularly at the lower educational levels. Thus, the principal criterion is policy, rather than financial resources.

Libraries/information centres may wish to be co-operative, but conflict with their policies can cause difficulties. The established policy in providing stock and services will be determined partly by factors which are to some extent beyond immediate control. These include the quality and quantity of staff, inter-library co-operation, "market pressures" from users and financial restrictions.

Evaluating the adequacy of the chosen library/information centre should follow the choice of educational/instructional objectives. A detailed evaluation is described in Sub-section 3.4.1. In the choice of library/information centres or their preliminary assessment a subjective judgment can be made of two factors:

1. <u>Pertinence factor</u>. This is a measure of how well the stock and services of a library reflect the needs and interests of the users. A very large national library will have a high pertinence factor because it acquires a wide range of literature. A small collection is dependent on skill in acquiring literature and on the extent of supporting services such as inter-library loans.
2. <u>Frustration factor</u>. This is a measure of the effort required to obtain reading material and information. A high pertinence in stock and inter-library lending is necessary, but skill in indexing, cataloguing and classification is also important where users are expected to be self-sufficient. Instruction may then have to be more in "mini-librarianship" than in information science. Reader and information services can significantly reduce the frustration factor by counter-balancing deficiencies and dealing with unusual cases.

Pertinence and frustration factors are theoretical considerations on which practical criteria will be based. Unfortunately, the practical criteria for education will be determined to some extent by the approach used in the lessons. Thus the library's catalogue can be the major finding tool or merely one of many finding tools.

In the practical evaluation of a library/information centre, one can consider a system's approach with the elements shown in Figure 2.2.3.

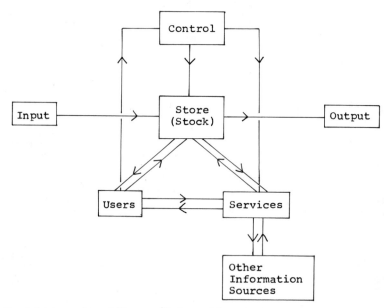

Figure 2.2.3 Evaluation of a Library/Information Centre

The inter-relationship of these aspects is given in more detail in Section 3.4.1.

2.2.4 Motivation from the Agency

Three kinds of agency have to be considered: the tutors responsible for undergraduate and postgraduate courses and syllabi in science and technology; employers of practising scientists and technologists; and the appropriate personnel in governing bodies of organizations who might sponsor or fund courses.

<u>Tutors</u>. These are the people responsible for determining the nature of the education of their students and for participating in their teaching. The reactions of tutors may represent an overall view of their departments or individual views. As far as motivation is concerned they can be put into three groups:

1. Tutors who readily accept information-retrieval skills as significant to their students' interests. Some of them will actively seek inclusion of information retrieval in their courses; some will seek reassurance about the suitability of the content and approach.
2. Tutors who are initially non-committal. (Probably the greatest number of tutors are in this group.) They may not actively seek inclusion of information retrieval in their courses but are prepared to consider it seriously, and to try it out, especially if brief written details and a convincing argument can be given.
3. Tutors who are opposed to inclusion of information retrieval in their courses. Fortunately very few tutors fall in this group, although a number have doubts about the approaches which are often used. Opposition may be due to:

   a. the kind of instruction they have themselves received;
   b. a different interpretation of information retrieval.

If a significant number of tutors are non-committal or opposed, it is worth linking instruction with the provision of an information service for teaching staff. The quality of the service will then become known and most tutors will reconsider their positions, particularly if the teacher of information retrieval acquires a useful reputation.

Employers. Techniques for convincing employers that they should support courses in information retrieval will relate to existing attitudes. Personal contacts are important, as is finding the right person to approach in an employing organization. A training officer is a useful contact and will be prepared to give time and serious consideration to any training programme. In other organizations the main need is to convince the person who can authorize attendance on courses of instruction. Obviously a direct approach should be more productive.

Acceptable courses of instruction are likely to be those which give obvious and immediate benefits. This is particularly true for employers, who regard attendance as lost work and therefore a cost, quite apart from the course fee. With enlightened employers it may be advisable to approach individual scientists and technologists directly. This has the advantage of producing highly motivated and responsive course groups.

Self-employed scientists or technologists, such as consultants, may readily attend courses of instruction if they believe that useful facilities, services or skills will result.

Government. Government should be interpreted here in the widest sense, ie. the body or bodies responsible for policy in an established order.

The governing bodies are:

1. Boards of studies;
2. Professional institutions;
3. Government departments;
4. International organizations.

1. Boards of Studies. In colleges and some universities the broad syllabus may be established by a board of studies. The inclusion of information retrieval in studies is probably more difficult to establish at this level unless the teacher can lobby effectively. Once inclusion of a subject is made at this level, some of the procedures for giving instruction are much easier.

2. Professional Institutions. Professional institutions and bodies can influence situations. The criteria which motivate them are related to professional development. Initiation of interest among professional institutions will cause some educational establishments to consider assigning special responsibility to a member of their own staff.

3. Government Departments. Government departments may occasionally initiate instruction in information retrieval. More often they support information centres, and it may be necessary to persuade them to agree to instruction and provide funding.

4. International Organizations. International organizations are in much the same position as government departments except that their main authority is their ability to give financial support and encourage training in many countries either in national programmes or in international courses.

3. Arrangements to plan a course

In planning a training course the teacher needs to have the right students in the right place at the right time. Although this is both obvious and simple in theory, it needs attention to ensure that it actually happens in practice. Also the practical nature of information retrieval and its application after instruction must be considered. For students, this means holding discussions with those responsible for the overall education of the students and possibly with the students themselves. In the case of practitioners, it means giving consideration to sponsorship, promotion and facilities.

3.1 ARRANGEMENTS AT INSTITUTIONS OF HIGHER EDUCATION

Information retrieval and other basic skills such as numeracy or scientific writing can be applied to all areas of study and should not be taught in isolation from the students' subject courses. For this reason, the teacher should seek to co-operate with those responsible for the subject courses. If the teacher is a member of the institution's staff, even if employed by another department, colleagues in the institution should be involved in the planning of the information-retrieval course. When the teacher is visiting the institution, it may be more difficult - but also more important - to work closely with the full-time staff in the planning phase.

Planning should start with preparation of a synopsis of the lectures, together with details of the practical exercises, as an expansion of the educational objectives. This synopsis can be used as a basis of discussion with the subject-course tutor, whose needs, comments, and additions are likely to be pertinent, generally useful, and often easily accommodated.

This discussion should be extended to other matters than subject content. These can include the timing and duration of courses, the status of practical work and academic examination within the subject. Information retrieval may be treated as a compulsory element in certain departmental courses, but this is a matter for the course supervisors. Any assessment, however, should be made by the teacher of information retrieval.

One must expect different course tutors to react in dissimilar ways to offers of lectures on information retrieval. The circumstances and responses can vary significantly within one institution and at different times, and sometimes the responses can be discouraging. Two circumstances tend to favour the teacher of information retrieval: firstly, an early recognition of the need for such courses by some practising and teaching scientists and engineers within the institution or locality; and secondly, a gradual change of attitudes to what exists and a growing acceptance of information-retrieval skills.

It is likely that at least one subject department will permit its students to receive course teaching, and provided that this is done competently then others will be encouraged to give assent.

If the teachers of information retrieval also provide an information service, they have a further opportunity to show the value of their particular skills. Time works to their advantage in that, as the course proves its value, the effects will become known.

Where programmes of instruction in information retrieval are established, the subject departments - especially heads of departments and course tutors - will reconsider their course structures periodically and will consider including information retrieval. The teacher of information retrieval should co-operate in this and should co-ordinate his other activities with the departments as far as possible.

The place of information retrieval in the total syllabus is often re-examined when there is a policy of reducing the hours of direct instruction, a change in the head of department or course tutor, or the appointment of a departmental staff member whose specialization includes information retrieval. It can be disconcerting if a department then expresses firm views which contradict those established in the teaching of information retrieval. Fortunately other departments are likely to adopt less restrictive arguments and even the more restrictive ones will mellow in due course. Thus the established programmes of instruction in information retrieval are not fixed but change or fluctuate in specific circumstances, sometimes with progress but sometimes with regression.

The economic situation which exists at the time the course is planned can influence course tutors and others responsible for educational content. During a period of expansion new facilities and new approaches may offer extensive opportunities to students. Hence, course tutors may find themselves attempting to restrict lecture sessions and service teaching in order to accommodate interesting and high-value laboratory work. A period of financial constraint on education challenges the resourcefulness of the individual institutions. At a time of deteriorating staff/student ratios and restricted expenditure on laboratory equipment, education in information retrieval appears to be more welcome.

Even if demands on the library do increase during times of financial restraint, the temptation to offer a cheap substitute to teaching must be rejected. It is the educational objectives which are important in teaching information retrieval, and one has been led to believe that in education in general a balance of self-discovery and interaction with the teacher is important.

In short, one should be positive about one's own beliefs and proposals (preferably written in brief notes), but should be ready to meet any particular requirements of course supervisors.

3.1.1 Timing, Duration and Structure of Courses

Like many other aspects of teaching information retrieval, the timing, duration and structure of the course must reflect the particular circumstances. A modular programme can be used as a basis of discussion with departments.

Timing. The basic criteria are:

1. Early need for information-retrieval skills;
2. Changing needs at different times;
3. Association with immediate and extended needs.

Arguably, information-retrieval skills are needed as soon as one is able to read; but the skills then needed are simpler than the demands of self-learning and research that are associated with postgraduate studies.

Ideally, information retrieval, together with its associated topics, should be taught at different parts of the overall educational programme, for example, just prior to a time of intensive or extended need such as project work. But circumstances may limit the teacher to a single series of lessons. Careful planning, execution and follow-up are necessary to ensure effectiveness.

In practice, the choices are as follows:

1. As soon as the subject course or programme demands the use of literature beyond a basic reading list.
2. Immediately preceding substantial project work.
3. A compromise between the ideal and the possible. Course content relating to studies can be given early in the educational period; that associated with particular exercises can be given before the project work; and that related to practitioners' needs can be given at the end of the educational period, before students leave to start their careers. Some repetition or revision is desirable on the third occasion.

A further need is to avoid what may be termed 'weak' lecture periods ie. those at the end of a long series of lectures, at the end of the week or at the end of a term or semester.

Duration. The content of the lessons given in Part II of the Guide for use in institutions of higher education, particularly with undergraduates, is based on three one-hour lessons plus practical exercises. In extreme circumstances this schedule can be reduced to one hour of lessons supplemented by adequate handouts, although students may then be uncertain as to what is expected of them, unaware of the purposes of the course and unable to exploit the necessary skills in the future.

Any further aims and objectives will require more time than is suggested above. A series of lessons lasting one hour each week for a full term can be helpful in developing use of particular types of information such as statistics for economists or managers and product information for design engineers. Some courses have extended beyond ten hours' formal instruction plus appropriate exercises. When instruction extends to ten hours or more, steps should be taken to measure the overall effectiveness of the instruction and the performance of the students - especially in examinations, course work and vocational activity.

In practice, the teacher will normally suggest a suitable duration which can be modified after discussion with course tutors and after the initial series of lessons. It may be tempting to seek substantial programmes, but these can usually be given for only a few departments and normally teachers' responsibilities extend to all the departments. Cases may arise where a substantial programme of instruction is developed for the few enthusiastic and co-operative departments, and this then stimulates departments which are non-committal, opposed or not consulted.

3.1.2 Arrangements for students

These consist of:

1. Informing the students of the course by time-tables;
2. Inclusion of the course with associated topics;
3. Reinforcement of the course by department;
4. Location of the course.

Informing the students. A number of attempts at teaching library use or information retrieval have been disappointing because of very low initial attendance. During preparation one should ensure as far as possible that attendance is good at the initial session. It is then up to the teacher to capture the interest of his audience and thus maintain attendance throughout the course.

In practically all cases it is essential that information retrieval, and any other serious topic, is time-tabled. Some students regard timetables as 'the law' of education and arrive at every designated point at the designated time. Alternative methods can lead to disappointment, e.g. notices, some of which seem to be read by some of the people some of the time. Some students regard anything not time-tabled as being either insignificant or of no concern to them. Tutors, on the other hand, often organize tutorials or similar activities and thus take up apparently free periods. If the institution has an administrative assistant with responsibility for arranging the lecture programme and timetables, it is useful for him to ensure that there are no overlapping commitments and that audio-visual material, handouts, etc. are available as required. The teacher can then concentrate on the specific role of teaching, and not suffer continuous frustration as a result of minor administrative problems.

Another point is the use of an appropriate title on the timetable. From experience it seems that "Library Use" or "Library Lectures" may actually discourage attendance, while "Information Retrieval" encourages it. Attendance is also likely to drop off if the regular sequence of lectures is interrupted by postponement or a change in the time of one session.

Inclusion with Associated Topics. Information retrieval as a topic, particularly for scientists and engineers, does not compare fully with other time-tabled subjects. The duration is generally shorter, it is service teaching, and generally is a little out of context. Response may be improved if it is associated with similar topics, possibly under a generic title. "Communication" is one obvious choice, another is "Study and Work Techniques". For the mature student the topic may form part of a programme entitled "Return to Learning".

Reinforcement by Department. Sometimes poor attendance may be thought due to apathy, even lack of interest, by the department, which is reflected in student attitudes. In contrast, some departments have established the practical exercises as course work, followed the exercises by projects on which information-retrieval skills have been assessed, or being included as examination questions. Attendance and standards in such conditions have been found to be good.

Location of Course. This is dealt with in Section 3.4.

3.2 PROMOTION AND SPONSORSHIP OF SHORT COURSES FOR GROUPS OF PRACTITIONERS[1]

The reasons for holding a short course can affect the nature of promotion and sponsorship and may include:

1. The need to improve the ability of the participants to use existing information services already known to them.
2. The need to make them aware of other information resources and retrieval techniques.
3. The need to discharge the responsibility of the sponsoring organizations for in-service or continuing education.

The UNISIST guidelines may best be illustrated by the case of a visiting teacher assigned to a small technical information service of one or two people with minimal facilities but assisted financially and materially by educational, government and regional interests. An

---

[1] A complementary study issued in connexion with this Guide is concerned with short-course organization and should be consulted, together with the following comments which are based on experience with this specific type of course. This study is: "Guidelines for the Organization of Training Courses, Workshops and Seminars in Scientific and Technical Information and Documentation". Prepared by P.Atherton, Paris, Unesco, 1975. 88p. (Doc.SC/75/WS/29)

advisory committee or employing authority may well require the service to assist in improving productivity or performance of local users by stimulating enquiries and providing some in-service education by short course instruction. Such a case embodies most of the principal problems which may arise and it is wise to follow a tried and proven approach which has been used in similar circumstances.

One approach that has been used successfully is for the teacher to concentrate on the educational content and the information centre to deal with administrative matters. An initial discussion between them should clarify objectives, the programme and the tasks, which are as follows:

Task 1 - Specifying the Target Audience. The sponsoring agency should be better able to judge the potential audience. The teacher may ask for important constraints - on maximum of group size, number of courses, and preferred grouping (by subject, employer, level). See Task 4.

Task 2 - Specifying the Short Course Instructional Objectives. This is mainly the function of the teacher, bearing in mind the reason for the short course, and is subject to discussion with the sponsoring agency or the host organization.

Task 3 - Producing a Short Course Programme. This follows Task 2 and can be based on an existing programme such as that shown in Figure 3.2.

Task 4 - Establishing Constraints. These are the joint responsibility of the teacher and the sponsoring agency, and can be posed as questions:

a. What facilities are available in terms of lecture rooms, library resources and equipment (e.g. audio-visual), for lessons and exercises?
b. What is maximum size of group? Given ample facilities, twenty to twenty-five is a reasonable maximum if there are at least two people to serve as instructors in the practical sessions.
c. How many short courses are to run?
d. What finance is needed?

Finding Information: A One-Day Course for Engineers and Scientists

9.00	Lecture: Introduction to Information Searching
	A brief tape/slide presentation establishes the foundation of the subject and is followed by a lecture on the basic theory of information needs, storage and retrieval.
9.45	Tour of the Library
	An opportunity to see some items of interest and to get one's bearings for the practical work.
10.05	Practical Work I - Quick Reference Questions
10.45	Coffee break
11.00	Discussion
11.30	Lecture: Principles of Subject Searching
	A description of the techniques of locating information of value to the scientist and engineer.
12.30	Lunch break
14.00	Practical Work II - Subject Search
	Each course member is given a typical industrial situation and suggested information requirements. Information officers provide any required assistance.
15.00	Tea break
15.20	Practical Work II (continued)
16.30	Discussion
17.00	Course disperses

Figure 3.2 Example of a One-Day Course Programme

Task 5 - Obtain Facilities. Obtain agreement and dates for use. The facilities needed can be specified by the teacher and located by the host institution. They will include a room for lectures and discussions; a library which will have stock stipulated (and possibly supplemented) by the teacher; and any accommodation and meals that are necessary. These tasks are for the host institution.

Task 6 - Obtain Finance and Authorisation. The course may involve expenditure of cash, if only for meals and refreshment. The sponsoring organization should ensure that due provision is made. It may be able to meet low costs itself - ideally without having to spend time in obtaining authorization from its committees or the employing authority. Even if it is not required to meet costs, authorization may be needed if an instructional responsibility is not implicit in the organization's terms of reference. Charges can be avoided if the participants pay their own expenses and facilities are free. If there is a course fee to cover expenses, a minimum attendance may be necessary. Agreement on disposal of surplus income can be important in maintaining goodwill. Sponsorship from national and international organizations is often worth investigating.

Task 7 - Publicity. Publicity will be related to the target audience and best carried out by the host organization or the sponsoring agency. If the teacher has been involved in a number of similar courses, the organization will appreciate details of the relevant publicity.
Personal contacts are very effective, but supporting material such as posters and brochures will be needed. Announcements in the professional journals and direct mailing to individuals using professional mailing lists can be useful. Local newspapers can also be used.
If the potential audience is limited, e.g. to a particular employer or industry, then the individuals may appreciate details of the programme as well as a brief description of the short course.

Task 8 - Course Preparation. The teacher will need brief details of the participants, especially names and jobs. Qualifications and years of experience may also be useful. The sponsoring agency can best supply these details to the teacher who can use them as a basis for preparing the practical exercises. Sometimes this is a time-consuming task and may be difficult for a teacher remote from the region. If so, the host organization should provide assistance.

Task 9 - Presentation. Much depends on the teacher, who should be free to concentrate on making the best possible presentation - administration, welfare and practical matters being looked after by the host organization. If the host organization also provides an information service that can be used after the course is finished, it should take a leading role in the presentation because real benefits can accrue.

3.2.1 Promotion of Short Courses

The promotion of short courses is concerned with persuading sufficient members of the target group to attend. The actual method of promotion will depend on the target groups, which may be:

1. Senior scientists, technologists and managers: people in senior positions of responsibility. As a target group their reasons for attending may be to learn to recognize the role of information retrieval and of library and information services. One aim of a course for this group is to promote understanding of the value of information resources in relation to their cost. Promoting short courses for very senior people is probably the most difficult form of promotion and should not be attempted without experience of other short courses or without ample preparation and resources. Very high standards need to be set for practically all course elements, including visual aids. Such standards can be an additional expense and a high participation fee may be necessary. This has the added advantage of attracting only the most senior people. However, it may result in too small a group to be workable. In certain cases attendance by invitation can be an effective approach; but this implies adequate funds to finance the course.
2. Experienced scientists and technologists; these being well-established people in fairly responsible positions. Their understanding of information retrieval may need enlarging in view of current developments.
3. Research scientists. They need to be informed of new developments in their field. Also they may need to publish, so new developments in publishing should be included, as part of the information-transfer process, in a course for this group.
4. Industrial researchers. They need to know what is occurring in their field of interest, especially in the patent's area. Often what they need is not available through commercial channels.

5. Technologists, who are employed in factories and workshops. They need immediate solutions to problems and generally do not read or publish. Repackaged information may be their main requirement.
6. New practitioners who are entering the field. The short course may be an extension of their education, with the possibility of specific training for the new job. It can also establish the views of these people on information techniques.

3.3 TEACHING STAFF FOR THE COURSE

The ideal teacher for information retrieval would have:

a. knowledge of information retrieval,
b. subject knowledge,
c. teaching ability.

These characteristics are associated with librarians/information officers, scientists/ technologists and teachers respectively. It is unusual to find all three characteristics well developed in one person. In practice, the relative significance of each characteristic has to be considered.

A knowledge of information retrieval is obviously the most important characteristic. This knowledge is best developed in librarians (practising or teaching), particularly those involved in information work. An information officer can be considered as an extensive or well-developed user of libraries and information centres. Thus the first choice for teaching staff would seem to be a librarian/information officer, particularly one currently in practice who is familiar with the current situation. To be effective he must convey the information on skills rather than merely present it, hence the significance of subject knowledge and teaching ability.

Subject knowledge is important for developing an appropriate viewpoint, understanding the key needs of users, and maintaining a high relevance in the lesson content. Instruction from a professionally-trained and experienced scientist or technologist can provide significant benefit. However, the degree of specialization and the background of the people attending the course has to be considered. The teacher of information retrieval will usually find a general subject (as opposed to specialization) sufficient.

In the perfect situation the teaching would be done by an information officer attached to a group of scientists or technologists working in the relevant subject field. Such a person would be familiar with the range of relevant up-to-date information needs and retrieval skills. Illustrations and examples from the real cases could usefully enrich the lesson content. This perfect situation does not occur frequently. More commonly teaching is done by librarians and information officers with backgrounds in science and technology and subject teachers who are experienced users of libraries and information centres. The first of these two groups is very suitable because its members can meet the principal requirements. They may have lost their detailed subject knowledge, but the scientific background remains, and this provides the essential understanding to teach scientific users. Specialist information scientists are useful supplementary teachers for the very specialized aspects of information retrieval.

A scientist or technologist who is familiar with the techniques of information retrieval is also worth serious consideration. Such a person may lack breadth of knowledge relevant to information retrieval but will normally have a highly relevant directness of approach. A very satisfactory situation is co-operation in teaching between librarians/ information officers and knowledgeable users.

The third useful characteristic is teaching ability. Information retrieval is a complex subject - a major interface between the information needs of users and the information content of the documents held by libraries and information centres. To present the essentials in a meaningful, clear and relevant manner needs substantial teaching skill. Teaching skill may occur naturally in a few individuals and can be usefully developed in most people.

The agencies and libraries which are initiating courses need to find suitable and available teachers for the courses. A librarian/information officer with scientific training and teaching ability will be easily identified if he is available locally. If not, it may be possible to obtain the services of a visiting teacher, who will need teaching support from a person familiar with local services. Similarly, teaching teams in which individuals' skills complement each other can be developed according to the circumstances. Teaching teams will need either a team leader or an agreed division of responsibility.

3.4 LOCATION OF COURSE

Information-retrieval skills are practical skills which need the equivalent of a workshop - in a library or information centre. What is required of the workshop is often determined by the nature of the course. In many cases alternatives can be found when particular items are lacking in a collection and teachers may also be able to supplement the collection with their own material. Thus a particular library or information centre should be evaluated by checking the educational/instructional objectives and then interpreting them into realistic goals.

It is then desirable that the arrangement of the collection and any local variations which affect the instruction are explained to users attending the course. This may already be done to some extent by the library itself in guides, library tours and prepared presentations. (See Section 3.4.1 and also the details for institutes of higher education in Appendix 1.)

3.4.1 Evaluation of Library/Information Centre

For a thorough evaluation it is useful to analyse the total library or information system and to study each part as it relates to the whole.

Store (Stock) may be a centralized or a fragmented collection. A centralized store permits the most effective and efficient use. Its limitation is the inconvenience caused to some users who do not normally work near the library. Inconvenience can lead to frustration, but it can be reduced by certain service functions. Of these telefacsimile may eventually prove very useful, but costs and technical problems prohibit it at the moment.

A fragmented collection is desired by some users, but it can lead to serious deficiencies or can increase running and capital costs. For instruction in information retrieval a divided collection can be acceptable if the finding tools are available at a central point. A finding tool may be considered as a reference book, guide, bibliography or abstracting or indexing service which locates factual or bibliographic information. Bibliographic finding tools contain references to the primary literature of books, journals, papers, etc. Thus the location of Chemical Abstracts and Engineering Index will be important for chemical engineering work. If the finding tools are scattered, it may be possible to bring them together temporarily for practical work on a course. This is particularly useful if the tools are single-volume works such as compilations of statistics. However, experience shows that much time and effort can be spent in temporarily bringing selected works together when several collections are involved. If those attending the course are to continue to use the library or information centre in the future, any temporary relocation of material should be explained to them.

The next consideration is the relationship of finding tools to the primary literature of books and serials. An information centre may have a small collection of primary material and a high concentration of finding tools. In short, this collection is likely to be orientated very much towards finding information. A frequent problem of information centres is the high cost of some finding tools in fringe subject areas; but this appears to be a lesser problem for academic libraries.

If the quantity of finding tools is low, their quality is the next consideration. A few very good tools can suffice where quality relates to extent of coverage and quality of indexing and abstracts. Relevance of coverage such as subject or geographical scope can be an additional factor. Many libraries make compromises and obtain discarded copies of key finding tools, or purchase selective editions, e.g. one in three, of annual works like a handbook of chemistry and physics. Thus one function of evaluation is to check the acquisitions policy either by discussion with library staff or by sampling material.

Services. The services and stock of a library are interlinked. A national library may provide extensive stock with minimal services. A specialized information centre for a widespread clientele may have minimal stock but fast, reliable and extensive services. These two cases are extremes; usually there is a balance between stock and services according to circumstances. In some cases of financial restraint priority must go to stock, for without a certain amount even the basic provisions cannot be met. Also it can be easier to account for benefits in terms of assets held in stock than in terms of services to users.

One significant factor determining the relationship between stock and services is the amount of inter-library co-operation.

Users. The types of user and the relative numbers of each type, both actual and potential, must also be assessed.

The main types considered in this Guide are student users, who can be sub-divided into undergraduate, postgraduate (course) and postgraduate (research); and practitioners, sub-divided into scientists/technologists and other users of scientific and technical information.

Loan records can be a useful guide in evaluations: they indicate in particular the types of user that make very heavy use of the library. In an academic library undue prominence of undergraduate users may have indications different from undue prominence of research users. The same is true for, say, management users and technician users in an information centre.

Control lies with the chief librarian or information officer; he establishes policy, which is then implemented by the library/information staff. In exercising control he should take account of the needs of users and other relevant factors, and he may be accountable to a committee or group of users which, in many cases, will actually determine the policy of the institutions.

Input. Studying input is really a continuation of studying the stock. Changes in input quickly reflect changes in policy and can give a better understanding of how to evaluate a library/information centre. They are also one useful way of explaining a library - in structure and arrangement. Unfortunately it is not easy for people other than the library staff to examine the details of input.

The currency of the input - dates of publication and dates of ordering - should be considered. A proportion of discarded material from other libraries will indicate attempts to maintain the provision of literature in difficult circumstances, such as lack of funds or recent establishment of the library or of interest in the subject.

The accounting procedures used in payment for the material are a useful indication of how the control element is operating and what external influences may exist. An allocation to each subject or department may indicate strengths in specialist, rather than general material. Alternatively, it may be simply a monitoring operation designed to avoid major deficiencies in stock. A detailed breakdown of figures (if easily available) can indicate strengths and details of control. The relative costs of staff and stock can be particularly useful.

The way material is handled after acquisition is also useful information. Some difficulty may be experienced in determining the procedures for handling finding tools. These are often classified and catalogued differently and may be put on one side for special treatment. Particular finding tools such as book-lists, national bibliographies, printed catalogues and periodical directories are stocked for librarians' use in the same manner as publishers' catalogues. This may be efficient materials-handling but it is bad service provision.

Output. One can measure the number of loans, enquiries and similar criteria, but measurement of real benefits is more difficult. If an information centre keeps a case file of productive information service, it can be usefully studied. However, few enquiries for information lead to a very high financial return. Thus a study of five hundred enquiries may show negligible returns - or returns only slightly more than the cost of providing the information. Some of the return may, of course, be concealed.

Checklist of practical questions concerning the library/information centre

Service — how does it react to possibilities?
- increased loans
- increased enquiries
- increased demand for material in stock
- increased photocopying

— how able is it to cope with nominal increases?
- 5% increase over one month (visiting teacher)
- 15% sustained increase from continued policy of educating the user

Stock — what are the main collections and sub-collections?
e.g. 1. books - loan
- loan oversize
- loan pamphlet
- reference

2. serials

— what are the guides to the local collections, especially by subject?

Finding tools — where are they located?
- what logical order?
- number of items?

Currency — how recent are the editions of significant items such as handbooks?

3.4.2 The Classroom

The place of teaching and learning is significant. Ideally, one should visit it in advance to ascertain its possibilities and restrictions. In an unfamiliar lecture room it is easy to risk unwanted jokes when trying to cope with screens, lights, heating, curtains, seats, projectors and the other paraphernalia of the classroom.

For lessons an informal seating arrangement may be appropriate and the suitability of the room needs checking. So does its suitability for a film or a tape/slide presentation.

Regarding lessons on information retrieval there is some discussion on whether the better location is a lecture room or a room in the library. Although there are circumstances where one is more appropriate than the other, the experienced teacher will not regard the issue as important for all cases.

The library-based room is often sought by the inexperienced teacher. It is likely to be, or become, a more familiar place. Numerous finding tools can easily be transported there, and users have a sense of being close to the stock. This may be satisfactory for teaching library use. Students of information use may feel that the teaching is part and parcel of the syllabus if their normal lecture rooms are used. In this Guide the use of normal classrooms is advocated unless stock is actually used, in which case the lesson should take place within the library. This may create problems of containing noise and activities; best solution would be a library workshop - a room leading off from the library, ideally at a point adjacent to the collection of finding tools.

4. Preparation of courses

Preparation of lectures and practical work obviously takes a great deal of time and thought. The amount will vary depending on the teacher's experience, the amount of material already prepared, and special requirements for different groups of students. It has already been said that background information should be supplied on the students taking the course so that practical work can be related to their interests and abilities.

The degree to which the teacher is prepared to make changes during the course of the instruction should be considered. With small groups one may almost ignore all prepared lecture notes and allow changes in approach as important points arise. With very large groups one is likely to have prepared a full script - as in most other processes of mass communication. The normal situation usually lies between these extremes and the final choice is usually a modification of the prepared notes. Thus in each case it is necessary to weigh up the advantages and disadvantages of prepared presentation and a spontaneous style appropriate to the situation.

The best presentation takes account of as many local circumstances as possible and incorporates them in the brief notes, so that what looks like a reaction to a situation may have been prepared. Such an approach may be made by preparing a standard full script (ideally in typescript, with a large fount), together with cues in the margin. The teacher can then expand upon the headings and related topics during the lectures, whilst ensuring that sufficient time is given to the essentials of each point. If the lecture starts to become disorientated and without purpose the full script can be resorted to, so ensuring full and proper coverage in a clear and precise manner but losing spontaneity.

Chapters 5, 6 and 7 are aimed at providing material for a range of situations and may be used as the basis for full lecture notes. They are, however, intended for the ideal situation of an alert and responsive audience, accustomed to an environment of extensive facilities including libraries, and educated both academically and vocationally.

If the audience in a developing country can see little likelihood of applying immediately the information-retrieval skills being taught, the long-term advantages must be strongly emphasized. It is sometimes possible and indeed necessary to reorientate retrieval skills to meet local circumstances. The ideal procedures of this Guide can then be largely disregarded and individual users can be helped to achieve their immediate objectives. Chapter 9 includes some suggestions for compromise approaches which may be used when the local situation is far from the ideal.

To summarize, preparation of courses cannot be done without correlating the various aspects of objectives, variations in approach and nature of the audience. Feedback and evaluation permit useful modifications at the various stages. These aspects are discussed further in this chapter as pictured in Figure 4.0.

4.1	Detailed working objectives
4.2	Interpretation of the objectives
4.3	Variations in approach
4.4	Variations to reach different audiences
4.5	Variations for imposed limitations
4.6	Teaching and learning aids
4.7	Use of case studies
4.8	Feedback and evaluation

Figure 4.0 : Correlating Various Aspects of Objectives

4.1 DETAILED EDUCATIONAL/INSTRUCTIONAL OBJECTIVES

Further details of the objectives of user education in Section 2.1 are required - with particular reference to the different target groups. These details are given in the next three sections and are followed by a section on certain aspects of group attentiveness and interest which are applicable to all target groups.

The objectives present, in outline, what the course should cover. They may be covered in lectures, discussion sessions, and, in many cases, in the practical work planned for each course.

4.1.1 Undergraduate Courses

The following objectives should be used to stimulate thoughts on specific cases:

Objective 1. Students working individually should be able to locate sufficient relevant books and periodical articles to produce a substantial essay on any topic which is widely accepted as part of their subject or vocation.

Qualifications to Objective 1. Locating relevant books means using an academic library's catalogue and at least one relevant published booklist (e.g. the national bibliography or a list of books in print for the book trade).

Thought needs to be given to the quality of the available lists and the effect of faults in the library system. The tests or practical work should be based on straightforward answerable questions, without significant problems of terminology.

Locating periodical articles means the use of at least two abstracting journals, one of which should cover multiple subjects (e.g. Referativnyi Zhurnal, Bulletin Signalétique, Chemical Abstracts, Engineering Index) and one specific in coverage (e.g. Zinc Abstracts, Corrosion Abstracts).

"Substantial essay" is an essay expected to be produced over several weeks.

"Any topic which is widely accepted as part of their subject or vocation" is self-explanatory. It could be the subject of a lecture or private study, followed by an essay; the objective should relate to both cases. Thus a student may have been previously ignorant of the topic and the literature on that topic.

Reasoning of Objective 1. Student tasks include producing studies (essays) of particular topics and supplementing lessons by private study to prepare for examinations.

Objective 2. Students should be able to acquire data and other factual items of information which may be needed in the laboratory or workshop, in practical exercises and in their vocations, provided that the information is clearly documented and can be located by a simple and obvious search method in about five minutes.

Qualifications on Objective 2. "Data" means inherent physical or chemical properties of materials, listed in a limited number of specified reference books (five is a suitable number) with adequate indexes and with easy location by classification or listing in a subject index of the library's holdings.

"Factual information" means information for similar needs located in a similar manner to data. Examples are safe working loads, optimum power ratings, load-carrying potentials (conveyors and so forth), circuit diagrams, mathematical and other formulae, mathematical constants and conversion factors, material properties, heat treatments, flow rates and equilibrium diagrams (metallurgy).

"Specified reference books" are those with high use and potential value to the group in question.

Reasoning of Objective 2. Based on an understood need for data and similar information in studies and vocational work. The careful selection of reference books is a precaution to avoid giving a misleading impression of the relative value of different publications.

Objective 3. To be able to locate the names and addresses of organizations which are likely to provide supplementary information to published information on any topic widely accepted as part of a subject or vocation.

Qualifications to Objective 3. The topics will be chosen as listed in up to five reference books and without terminology problems.

The reference books will be chosen according to their usefulness and potential value.

The organizations concerned can be domestic (in the same country) or international.

Reasoning of Objective 3. Agreed opinion of many practitioners that the most valuable training is building a series of informal contacts, supported by the need to correlate recorded sources of information with other sources - classified as useful.

Objective 4. To be able to appreciate the terminology and interpretation problems of information retrieval, even when other factors are straightforward.

Qualifications to Objective 4. One or two major publications are used and questions are set in which terminology and interpretation are relevant.

A useful source is a statistical question based on a major yearbook of statistics such as United Nations' Statistical Yearbook.

Reasoning of Objective 4. Finding tools vary in quality and style. Also information needs can be specified in various ways. There is arguably a need to overcome these obstacles and appreciate the reasons for their existence.

The remaining objectives will depend on the various needs of each subject area, e.g. for such sources as standards, codes of practice and product information.

4.1.2 Postgraduate Courses

The following are the relevant objectives with this particular target group:

Objective 1. Students working individually should be able to locate books, periodicals and other published material relevant to a piece of research or study over a period of three months, one year or three years, according to the duration of their dissertation or thesis.

Qualifications to Objective 1. "Locating relevant books" means using the academic library's catalogue and relevant published booklists and bibliographies which are in stock. It can also mean using a more specialized or comprehensive library if this is accessible and appropriate to the course or research.

Thought needs to be given to the quality of the available lists and the effects of faults. The tests or practical work should be based on realistic situations and should include problems of terminology and uncertainty. The latter can be achieved by inclusion of at least one question believed to be unanswerable.

"Locating relevant periodical articles" means the use of relevant abstracting and indexing journals, provided they are available. It may include consulting a subject index of abstracting and indexing journals.

Reasoning of Objective 1. The survey of the literature is regarded by many educational authorities and teachers to be part of a dissertation or thesis submitted by a postgraduate student. The standards of the literature survey should be similar to those set for other aspects of the submitted work. Another, not necessarily different, requirement is that a postgraduate studying librarianship and a postgraduate studying, for example, physics should reach equally high levels of performance in literature searching. However, it is very important that their actual performances should reflect the differences in their viewpoints. One should expect a higher "recall" value from the library student and a higher "relevance" value from the physicist. (See Objective 4 for definitions). Furthermore, one should not expect a physicist to equal a science-trained librarian in performance in a scientific literature search outside physics.

Objective 2. Postgraduate students should be able to acquire data and factual items of information which could be needed in the laboratory, workshop, practical exercises and vocational work, provided that the information is accessible in reference books, abstracting journals and pertinent organizations (apart from the more obscure sources). In addition, the student should be able to apply notional limits of applicability and confidence to the values according to the authority of the sources and procedures for evaluation and compilation.

Qualifications to Objective 2. These are mainly subjective limits on what is reasonable according to the potential value of the information and the effort required to obtain it. The books or abstracting journals should themselves be available and locatable through a logical index or arrangement. Within the books there should be satisfactory indexes or arrangement of material. If the source of information is an organization, it should be prepared to give information, and the name and address should be available through directories.

Limits to applicability and confidence need to be considered by both the teacher and the student. The teacher has to consider the needs of the vocations, available teaching time and the relative importance of various topics. The student has to relate the limits to the potential value assigned to particular pieces of information and to the effort needed to acquire them.

Reasoning of Objective 2. Based on the understood need for data and similar information in studies and vocations. The real problems need stressing to avoid giving misleading impressions on what data is available, the relative ease of access, and their acceptability (including limits both to applicability and confidence).

Objective 3. To be able to locate the names and addresses of organizations which are prepared to give supplementary information relative to a project or vocation.

Qualifications to Objective 3. The level of acceptable performance will be related to the availability and quality of directories, i.e. to the effort required to locate the information and the potential value of information of this type.

Reasoning of Objective 3. The literature cannot satisfy all information needs, but it can indicate potential unpublished sources of information. In reality, this objective is the first of several which relate to specialized categories of information and literature.

Objective 4. To be able to appreciate key variables in seeking information:

1. The role and potential value of scientific and technical information.

2. The effort, particularly time and frustration, in seeking information and relating this to its potential use and value in specific cases.

3. The occurrence and identification of information needs.

4. The concepts described by information scientists as:

Relevance - the measure of the closeness of a retrieved piece of information to the specified information need (enquiry).
Recall - the measure of number of pieces of information retrieved that are relevant to the enquiry in terms of the total number of relevant pieces of information in the store.
Classification - a scheme or outline of knowledge displaying mutually exclusive and collectively exhaustive categories or topics. Hierarchical classification suggests the positioning of various words in a chain going from the general to the specific.
Indexing - the process of analysing the information content of recorded knowledge and expressing this information content in the language of the indexing system.
Controlled vocabulary - a special list of words used in indexing with rules prescribed for selecting words and adding new words. Synonyms and various word forms (plural v. singular) are controlled by selection of a preferred form with cross references from non-preferred forms.
Free indexing - index terms are not controlled but assigned as each document is entered into the file, usually by selecting words that occur in the title, abstract or text of the document.
Natural language - the use of words selected from the text of the title, abstract of full text to provide index entry points.
Cross references - a direction (such as "see" and "see also") from one heading to another.
Multiple entry - the indexing or otherwise recording a document under more than one main heading.
Evaluated information - information that has been judged against other values to give some measure of reliability and accuracy.
Search strategy - the plan for retrieving information from one or more stores; it may be a list of subject headings to be searched or a logical statement for computer processing.
Indicative abstract - an abbreviated, accurate representation of a document which is descriptive in nature and states what the primary document is about.
Informative abstract - an abbreviated, accurate representation of a document which contains qualitative and/or quantitative information contained in the primary document.

Qualifications to Objective 4. The emphasis should be on teaching science and technology, not information science. The student, therefore, may need only to appreciate the existence of these concepts, not to study them in the depth usually prescribed in information science.

Reasoning of Objective 4. Based on the need to determine a suitable search strategy and to overcome possible causes of difficulty.

Objective 5. To appreciate the relevant special categories of information and literature applicable to the subject field.

Qualification to Objective 5. A necessary pre-requisite will be Objectives 1 to 3.

Reasoning of Objective 5. The understood information needs of the subject relevant to a field of study or vocation.

4.1.3 Courses for Groups of Practitioners

The following are the objectives to be achieved:

Objective 1. To increase productivity and reduce the problems and costs of industrial and commercial enterprises by making more effective use of information resources in the company or the region.

Qualifications to Objective 1. This is a broad objective rather than a specific instructional objective, but it deserves attention because it can be sought through two closely allied approaches. One is based on improving an information service and the other on improving the self-sufficiency of individuals.

Reasoning of Objective 1. The belief that the benefits of more effective use of available information can be significantly increased.

Objective 2. Practitioners should be aware of the recent developments and innovations which directly affect their work.

Qualifications to Objective 2. For course purposes the information needs to be published unless it is available through sources such as exhibitions and conferences.

Reasoning of Objective 2. Lack of awareness of recent developments is a common experience - with clear implications.

Objective 3. Practitioners should be able to carry out a feasibility study, or that part of a feasibility study which involves a survey of existing published knowledge on the topic.

Qualifications to Objective 3. Availability and accessibility of suitable library and information services.

Reasoning of Objective 3. Feasibility studies or surveys of the literature are believed to be a pre-requisite of a new research or development project.

Objective 4. Practitioners should be able to acquire data and factual items of information which may be needed in the laboratory or workshop, or in any other aspect of their vocation. In addition, they should be able to apply notional limits of applicability and confidence to the values.

Qualifications to Objective 4. Availability and accessibility of suitable library and information services.

Reasoning of Objective 4. Based on the understood need for data and similar information in the relevant vocation.

Objective 5. To be able to locate organizations which are prepared to give information relative to the field of work.

Qualifications to Objective 5. The character of organizations and quality of the directories need to be considered.

Reasoning of Objective 5. It may be more economical or effective to use an information service; when it is not, other organizations are often needed to provide supplementary information.

Objective 6. To be able to appreciate key variables in seeking information:

1. The role and potential value of scientific and technical information.

2. Management factors, such as identification and occurrence of information needs, evaluation of specific needs relative to time and effort needed, and control factors, e.g. extent of information needed.

3. The relative merits of using information services and searching for information oneself.

4. The obstacles and limiting factors, particularly of terminology.

Qualifications to Objective 6. A practical and vocational viewpoint is important unless the group is particularly intellectual and academic in its own viewpoint.

Reasoning of Objective 6. Based on the need to determine a suitable search strategy and means to overcome possible causes of difficulty.

Objective 7. To appreciate the relevant special categories of information and literature applicable to a vocation.

<u>Qualifications to Objective 7</u>. The wording of the objective is identical to that of Objective 5 for postgraduate students. There is likely to be a significant difference between the two groups because practitioners often have information needs beyond their subject field. These relate to their particular duties and may include a need for statistics, legal statements or official publications.

<u>Reasoning of Objective 7</u>. The understood information need in the vocation.

4.1.4 <u>Group Attentiveness and Interest</u>

An audience or group may or may not be motivated to assimilate the instruction. Certainly motivation cannot be expected, and cannot always be predicted. The relationship of information-retrieval skills to a field of study or vocation is not widely recognized. Thus, one should build into the course planning, and later the content, an element which, as far as possible, will maintain the attention and interest of the group.

The presence of existing motivation is usually a surprise, and an enthusiastic and keen group can readily achieve the educational/instructional objectives. Ideally, one should have an alternative lesson prepared in reserve for highly motivated groups (together with a second alternative for very unresponsive groups), and simple checks should be made to identify the degree of motivation as early as possible.

For student groups the lessons are normally spread over several days or weeks, and a group's personality becomes established and understood by their tutors.

<u>Undergraduate Groups</u>. These contain a range of abilities and may be receptive to new ideas without showing it. This is particularly so when they tend to accept lessons as a matter of routine. A confident and experienced teacher can stimulate an audience by deliberately breaking down an established routine and substituting novel approaches. There is a risk of confusing a group if this is introduced incorrectly. The topic and teacher are usually new to the audience, and undue novelty may result in students not taking seriously anything in the content. As in many cases, the key to success is a proper balance in using novelty of approach and familiarity of illustrations and examples.

In practice most undergraduate groups are unresponsive in the lessons. The exceptions usually result from development of communication skills earlier in their instruction, particularly if group participation or any form of group feedback has been encouraged. If a group is unresponsive, the lessons are likely to follow a prepared pattern, and it is difficult or impossible to relate events in the lessons to the quality of the practical work. When giving lessons to a number of groups of undergraduates it is natural and useful to compare group performances. However, their performances in practical exercises cannot usually be predicted from the events in the lessons.

In general, motivation in an undergraduate group is assisted by:

1. Clear support for instruction in information retrieval from their department, time-tabled lessons, embodiment of the course in their syllabus, evaluation of practical work/information-retrieval skills.

2. Balanced time-table, avoiding undue concentration of lessons or conflict with evaluation by project or examination.

3. Balance of lessons both between lectures and practical sessions and in style. Style need not be flippant or theatrical, although both may have their place. Teachers should make ample use of simple teaching aids, especially if they can be easily used and are of good quality. The audience may be accustomed to such aids and may expect them. Other groups, for which they are a novelty, respond particularly well.

4. Case studies and examples used as illustrations should stimulate by having intrinsic interest. In practice the examples may be news-worthy (e.g. space technology) or amusing such as the illustration on aeroplane design in Chapter 5 (Figure 5.1.2b).

5. Factual material should be presented in a readily acceptable form; handouts can be useful.

6. Practical exercises should be given as much attention as possible. They need to be stimulating, interesting and relevant. Ideally, they should use different approaches and should be graded in difficulty so that some answers can be achieved relatively easily.

<u>Postgraduate Groups</u>. These have similarities with undergraduate groups, but they are usually smaller in size, more mature and more responsive.

In comparison with undergraduate groups the following factors can be considered.

1. Support for instruction in information retrieval from the department is useful but not as essential as for undergraduates. Members of the teaching staff sometimes sit in with the postgraduates and this is sometimes useful, depending on the relationship between students and staff.

2. Time-tabling is still important, to avoid conflict of lessons.

3. Balance of approach and viewpoint are important. Information retrieval is common to librarianship and each subject field. Much of the terminology has been developed from librarianship and information science. Fortunately, some of these terms are taken from other sciences, for example 'noise'. Some have particularly scientific meanings - e.g. recall, relevance, precision. The viewpoint should as far as possible be that of a scientist or technologist, the specialist terms of librarianship and information science being introduced and explained where necessary.

4. Case studies and examples should be familiar and recognizable as arising from common experience. An undergraduate may prefer space technology, a postgraduate will respond better to the cases of laboratory experiments.

5. Factual material should be presented in a readily acceptable form and quickly considered. Discussion is usually more appropriate; handouts can be read later.

6. Practical exercises are best if realistic and of different degrees of difficulty.

Groups of Practitioners. These may consist of established groups or individuals who are together for the first time.

A well-established group can be handled in much the same manner as postgraduates, except that the spread of vocations may be much wider and qualification (examination and evaluation) may not be a common goal.

A group may be formed for a short course of which one topic is information-retrieval skills. The ideal arrangement in this case is for a skilled and experienced educator to take the first session on a broad and generally accepted subject. The teacher of information retrieval can sit in on this initial session and thus gain help in judging a suitable approach and content for his own contribution.

Short-course groups are the most unpredictable, and among them motivation may already exist. If it does, the teacher should start with a common interest as a theme and let it develop until it gathers its own momentum. Enthusiasm is most often found in groups in which individuals are attending on their own initiative.

Groups which come together for information-retrieval instruction only, need particular attention because of the problem of correctly assessing their abilities and interests, and producing a satisfactory content and approach. Useful approaches include high practical content, producing alternative approaches and 'free time' when the course assembles. In general, motivation of practitioners is assisted by:

1. Having a colleague look after administrative details and take part in the supervised practical sessions. This allows the teacher to concentrate on teaching, including motivation, and can make time available when it is needed most - for a last-minute changes during practical work.

2. Producing a clear outline of programme and objectives. Groups should know what is expected of them and should be able to rate their achievements.

4.2 INTERPRETATION OF THE OBJECTIVES

The broad objectives of instruction in information retrieval and library use are fairly easy to establish. In spite of differences in emphasis, it is likely that there will be some similarity in most of these objectives, though they are determined by different people. Effective use of the library; ability to locate information for studies and vocation; ability to carry out a literature search in specified situations - all these may appear in the objectives. Diversity appears in the interpretations of these objectives because different emphases are possible.

Closer examination of the overall objectives will normally reveal general statements which need interpretation into a series of defined goals, in relation to which an acceptable level of performance has to be decided. These defined goals, with clear measurement, are important in linking the definition of objectives and the actual teaching. One can inspire and instruct in lessons but this is wasteful if the benefits do not justify the time and effort. For example, an individual may develop skills in cataloguing, but without gaining any advantage.

Ideally the interpretations of the objectives (defined goals) should be established for common use, but this is not possible because they depend so much on personal approach and on existing knowledge. At best, they need qualification. The tutor should list his or her own goals together with a designated level of proficiency. The following list is intended as an illustration:

> Understanding the value of information
> Recognizing information needs
> Knowledge of the principal sources of information
> Knowledge that recorded information has limitations and obstacles
> Appreciation of finding tools
> Appreciation of search techniques
> Ability to locate introductory publications
> Ability to locate other literature and information

4.2.1 Understanding the Value of Information

The objective is to develop students' information-retrieval skills. The benefits of those skills accrue when the end product - information - is used to advantage. This implies that information has value and the students need to appreciate this fact. Unfortunately, it is very difficult to place quantitative values on many specific pieces of information. Thus only a general understanding can be gained, through case studies, of the real and potential contribution of information retrieval.

The tutor may be satisfied if students can state that information has value and a role to play in serious human endeavours. But this does not necessarily mean that the student has absorbed the knowledge or that he even believes it. Belief in the value or role of information in human endeavours is the first important goal. There is ample evidence in enquiries into disasters that awareness of available information could have prevented the disaster. Similarly, the history of successful projects and people often shows a useful role played by information.

If the lesson time is generous, more time can usefully be spent on this goal by requiring the students to study the role of information in particular cases and to compare findings. This will create a clearer understanding of the value of information which is so difficult to quantify. Although this study takes time, it yields additional benefits in understanding the scientific and technological processes, and information retrieval is more easily seen as relating to such processes.

4.2.2 Recognizing Information Needs

This goal is closely associated with the previous one - although it is less well-defined - in that an acceptable level of performance requires a statement with supporting illustrations. Such illustrations lead to the question, "How can individuals foresee the need for information?" An acceptable answer should be argued on the basis of common-sense and authoritative opinion until more detailed studies have been made of the area. Thus, 'novelty element' is one arguable case and 'departure from designated or routine practice' is another. An enumeration of the circumstances in which information needs arise may be sufficient, provided it is qualified by a note recording the lack of study in this area.

4.2.3 Knowledge of the Principal Sources of Information

In this area it is probably desirable to place the library, literature, or better still recorded information, in the context of the other principal sources of information.

In practice the ability to note recorded sources of information and the majority of the other sources is probably sufficient in a short course. In a longer course more time can be spent in developing knowledge on how the principal sources complement each other and on simple information needs which illustrate that one source may be more obvious in one case than in another.

4.2.4 Knowledge that Recorded Information has Limitations and Obstacles

If information-retrieval skills are to be used effectively then knowledge is needed of the limitations of the working material, that is, the documents and the other records. The extent and depth of knowledge required or considered satisfactory is vague and may be difficult to determine. The 'literature explosion' is a widely-used illustration of one obstacle. It is probably worth making some record of the other obstacles, especially the very important linguistic obstacles.

These four initial goals are not difficult to achieve and students can fairly easily make sufficient progress towards each one. The common element of the four goals is that they are not goals relating to skills but are rather "context knowledge" or an awareness of the use of the skills. Before the skill-goals can be considered, some thought may be appropriate to a series of goals associated with the logic of the skills, so that individuals can adapt and develop their skills according to the circumstances.

4.2.5 Appreciation of the Existence of Secondary and Tertiary Publications

One normally thinks of a publication as a document which is read from beginning to end for education, interest, pleasure or other reasons. Some books do not lend themselves to reading in this manner but are rather for consultation on a specific item. A dictionary is a classic example of such a publication. The boundary between the two types is indistinct and some users will seek facts from a textbook or document intended for full reading and others will read, rather than consult, reference works.

Among the reference works are many useful guides and lists which enable one to locate relevant reading material. Sometimes these finding tools are compilations of detailed information, containing brief items such as data and statistics; they may give synopses or indicative summaries. Abstracting and indexing services provide brief informative or indicative information on the full texts of books, journal articles, papers and other "primary" documents. For the most part these finding tools, are regarded as "secondary" publications, along with review publications. Lists of finding tools and similar works are usually regarded as a "tertiary" publication.

Finding tools have a key role in efficient literature searching and are the means of making effective use of selected items in the huge bulk of available primary literature. Adequate time and attention should be given to making clear the role of these finding tools and the different types of tools. It is the understanding of their role and the ability to locate pertinent tools which is important, not the committal to memory of specific tools - which may be superseded in due course. Satisfactory performance in achieving this goal can be measured in the ability to locate pertinent finding tools in specific cases.

4.2.6 Appreciation of Search Techniques

This is probably the most indistinct of the goals, which makes it difficult to determine an acceptable level of ability in students. In the suggested lesson contents in this Guide it is intended that the goal should be achieved indirectly by successful attainment of skill-goals.

If a generous amount of time is available, particularly with high-quality students, then the search techniques can be analysed and study made of the relevant facets. Examples of possible facets are: search procedure for books, periodical articles, reports and others; choice of search tools; choice of search terms.

Appreciation of the two types of facet - tools and techniques - leads to the consideration of their effective use to locate information. These skill-goals are set as practical exercises and measurement of achievement is fairly straightforward. Adequate thought and time are needed to ensure that appropriate goals are selected and that reasonable levels of performance can be expected. The actual goals are likely to reflect the personal beliefs of the tutor, who will need to select the goals and determine appropriate levels of performance.

4.2.7 Ability to Locate Introductory Publications (Books)

The value of this skill is obvious and need not be described. Some guidelines can be given orally or on paper, and this is useful because of the difficulty in using precise and clearly understood terminology in searches without asking for specific books.

An acceptable level of performance for university undergraduates should include locating relevant books in the university's library and also the recent publications of the home country. These aspects will need to be qualified by adequate cataloguing or indexing in the principal lists. The tutor needs to choose the appropriate tools and topics.

4.2.8 Ability to Locate Other Literature and Information

As in 4.2.7, this group of goals is of established value. It relates to the remaining significant categories of literature - notably periodical articles - and the appropriate specialized information needs. The lecture material in this Guide (Chapters 5, 6 and 7) provides some indication of the goals themselves, and appropriate levels of performance can be easily determined for local circumstances.

4.3 VARIATIONS IN APPROACH

The distinction between the use and value of different parts of the literature gives rise to a number of different teaching approaches. The two basic approaches may be referred to as the literature or librarianship approach and the information approach. The former starts from the argument that literature is a valued resource and that students need to understand how to exploit it. In certain cases it is undoubtedly the better approach because attention is directed to exploiting what is available. It is particularly useful in circumstances where needs are extensive, and in depth, and when the use of the library is readily accepted as a necessary part of study and practice. The student of history or culture is likely to be easily literature-orientated and to appreciate this approach.

In some cases a category of literature and an information need are closely linked, as in the case of codes of practice. An information approach can then be adopted throughout, and is sometimes preferable; for example with technicians who are very objective but do not wish for detailed background. The situation is comparable with instruction in servicing motor vehicles. The technician is concerned with maintaining vehicles rather than understanding all the problems of designing them.

In many cases a combination of the two approaches is the best solution. Thus, it is valuable to consider data as an information need and consider its various sources by an information approach. In contrast, most people will know when patent specifications are to be exploited and they can consider them by a literature approach. A scientist or technologist will normally accept the combined approach provided that its duality is clear. An acceptable starting or focal point is the interface between information needs and available literature.

There are other approaches that can be adopted or embodied in the instruction. One is the information-management approach. This is not really a separate approach but an extension or logical balancing of the main approaches. It is comparable with man management, materials management or energy management. It accepts information as existing and examines the need, use, control, benefits and costs of information. This is an acceptable approach but does embody additional aspects, such as management information systems and the responsibilities of individuals in reporting information.

4.3.1 Audience Attitudes

Some people do not intuitively relate the use of libraries to professional activities such as those carried out by scientists and technologists. Where this relationship is clearly seen one can adopt the literature approach, but in other cases the information approach is more convincing. In the more difficult cases audiences can be made more attentive by increasing the proportion of practical work, discussing individual problems and using smaller groups, preferably with tutorial and seminar techniques.

In cases of the opposite kind, where the audience immediately grasps the concepts or even predicts them, the real danger is in being unprepared for its responsiveness. The best approach is to present the background and theory as readily as the audience can digest it, with proper attention to feedback and participation in order to check their understanding, and then to use whatever time remains on realistic exercises - with supervision and guidance as required.

However, unresponsive audiences are more usual than attentive or eager ones. This may result from pre-occupation with more immediate problems such as examinations or tiredness. The best solution is to avoid difficult times for courses and to seek appropriate occasions, such as the start of projects, when information-retrieval skills can contribute to overcoming immediate problems.

4.3.2 Nature of the Student Groups

It is impossible to anticipate all the variations in the characteristics of student groups. One should, however, be as well-prepared as possible, and ready to modify course material as required.

During the initial preparation the numbers involved, the educational level, and similar characteristics can easily be ascertained. If there is a substantial degree of novelty in the course, it may be worth sitting in on some other lectures and discussing general aspects of them with the tutors. Copies of the timetable can be useful background as they indicate previous lectures and topics of study. The latter may be useful for preparing practical information-retrieval exercises. If student numbers are small, there may be sheets listing their names and basic details of experience. These can be useful in preparing material, particularly when student participation is planned.

One idea that may be tried with small groups of postgraduates is to request topics to be used as a basis for practical exercises. This in some cases develops motivation; but it has disadvantages, one being the limited amount of time available for preparing exercises, which can result in some questions not being directly relevant. Not all suggested topics may be suitable. Questions can be produced which are related to the agreed topics and the resultant exercises can then arouse interest without detracting from the educational objectives.

4.3.3 Availability of Resources

A well thought out lesson may produce substantial and satisfying results on one occasion, yet on another may apparently fail to stimulate or interest the audience. This fact indicates that careful preparation from reasoned objectives is sometimes insufficient and confirms that at some stage one must take local circumstances into account.

The most significant of these local circumstances is the availability of literature and other information resources. (See Section 3.4). Of lesser significance, but nevertheless troublesome, are the teaching facilities such as audio-visual aids. Comfortable seats, good heating and ventilation and other similar features can also be significant when one is attempting to stimulate and retain the attention of an audience which is tired or unresponsive.

The structure, stock and location of the library are all important and should be the first point of attention by the visiting lecturer, as discussed in Section 3.4. The effects in the practical exercises need particular consideration.

It is suggested that practical exercises are prepared before detailed lesson notes because of the time required to prepare reasonable exercise sheets. There is in fact an optimum period. Preparation too late produces hasty exercises and interferes with detailed lesson notes. Preparation too early does not cater for stock changes and lesson content. The actual optimum period will vary according to commitments and group sizes, but a reasonable general guide is around one month and not later than one week before they are needed. Visiting lecturers may not have that amount of time at the location before the initial lessons. In this case, a certain amount of preparation may be done prior to the visit - for example, a stock of standard question sheets ready typed up, or blank sheets (Figure 4.3.3) and actual questions on cards ready for typing on to the sheets. In either case, a checklist of finding tools or sources will be needed. An alternative course of action is to seek details of finding tools held in stock prior to the visit.

The practical exercises cannot be done in isolation since they would probably take too long to complete. Supervision during the exercises is the ideal solution but is not always possible because of the numbers of students involved. An alternative is briefing and discussion before and after the exercises. In this case, it may be useful to have copies of certain finding tools to hand.

The practical exercises are an essential part of the programme but they cannot cover all circumstances and variations. These variables may be covered in more detail if time allows, but they do require a considerable amount of preparatory work by the teacher.

The available resources and services may affect the teaching material in another way. For example, if a particular country maintains a standards organization which will assist with problems of standards, this resource needs to be covered in the lessons. The simple cases can be described without difficulty and reference can be made to the standards body in complicated cases. The visiting lecturer has to discover the local circumstances and should brief himself as fully as possible beforehand and on arrival.

4.4 TEACHING AND LEARNING AIDS

One cannot assume that motivation is present for learning information-retrieval skills, and that the subject content itself will capture or maintain interest. Teaching aids, with stimulating means of presentation, may be a useful supplement in attracting attention or maintaining interest; but there is a risk of using them as substitutes for basic teaching methods.

4.4.1 Handouts

Handouts are useful in providing permanent details for all lessons, but may not be timely for some information needs. They also permit students to give their full attention to the teaching, rather than record details in notes etc.

It may be unwise to spend a great deal of effort on an all-embracing handout, but better to produce a guide-book which can be easily updated with changes in holdings and locations.

For example, reference books often have to be selected for a particular enquiry from a collection of several thousand. Normal cataloguing and classification procedures give some assistance but usually not enough, and possibly 20 per cent or more are being continuously replaced by new editions. A more detailed subject guide can be produced which should be more precise than the catalogue and more up-to-date than any handout.

Lists of abstracting journals or national bibliographies usually have a longer life and can be better justified than simple lists of useful reference books. Those reference works that are not continuously replaced, such as Beilstein or Touloukian (Properties of Materials), should be treated as special cases.

4.4.2 Overhead - Projector Transparencies

The overhead projector permits the use of transparencies to provide good-quality lecture illustrations and to indicate salient points. They can often be hand-written - provided they are clearly legible - and are relatively quick and easy to produce. Their use is not really different in information retrieval from that in any other subject. The temptation to use them for detailed notes should be strongly resisted.

PRACTICAL LITERATURE SEARCHING	
NAME	COURSE ....
QUESTION	ANSWER
1. Trace an introductory text on	AUTHOR: TITLE: PUBLISHER: DATE OF PUBLICATION: SOURCE OF INFORMATION (& page):
2. Find the bibliographical details of a book entitled:	AUTHOR: PUBLISHER: DATE OF PUBLICATION: SOURCE OF INFORMATION (& page):
3. Trace a recent article on:	TITLE OR ARTICLE: AUTHOR: PERIODICAL TITLE: VOL.NO: ISSUE NO: DATE: PAGES: SOURCE OF INFORMATION (& page):
4. Trace a recent article on:	TITLE OF ARTICLE: AUTHOR: PERIODICAL TITLE: VOL.NO: ISSUE NO: DATE: PAGES: SOURCE OF INFORMATION (& page):
5.	 SOURCE OF INFORMATION (& page):
6. Give details of a recent article on	TITLE OF ARTICLE: AUTHOR: PERIODICAL TITLE: VOL.NO: ISSUE NO: DATE: PAGES: SOURCE OF INFORMATION (& page):
7. Give details of a recent article on	TITLE OF ARTICLE: AUTHOR: PERIODICAL TITLE: VOL.NO: ISSUE NO: DATE: PAGES: SOURCE OF INFORMATION (& page):
8. Give the name and address of a manufacturer of	 SOURCE OF INFORMATION (& page):
9. Give the number and date of the British Standard Specification for	 SOURCE OF INFORMATION (& page):
10.	 SOURCE OF INFORMATION (& page):

Figure 4.3.3 : Example of Standard Question Sheet

4.4.3 Photographic Slides

Pictures of better quality are possible with photographic slides than with overhead-projector transparencies. Their satisfactory use demands proper production, a darkened room for projection, and an adequate projector. If these criteria are not met the result is likely to be inferior to the more easily produced overhead-projector transparencies, which can be used under bright lighting.

Proper production usually requires the assistance of a photographic expert and careful preparation of the graphics. Photographs often emphasize difference in paper quality. Creating a darkened room ought to be fairly easy but it is not always so. If slides are to be used in an unfamiliar room, it is particularly desirable to visit the room before preparing the lecture in detail. Back-projection units, in which a projector is placed in a cabinet and behind a frosted glass screen, can be used with smaller groups in rooms which cannot be darkened. If students need to take notes, the amount of light may need adjusting several times during the lecture period.

4.4.4 Tape/Slide Presentations

Sound recorded on magnetic tapes is synchronized with the showing of a series of slides. A second track on the tape usually records pulses which change slides in the projector at the appropriate moments.

These presentations require setting up. Ideally there should be a separate operator in the case of large groups so that technical difficulties do not distract the teacher. For smaller groups, it is normally sufficient to set up the equipment and test it five to ten minutes prior to the lesson.

Tape/slide bits provide a presentation almost complete in itself. However, it is important that they are regarded as a supplement to the teaching rather than as a direct substitute. This is because they are more or less inflexible and cannot be modified for particular situations. Many lectures are of one hour's duration and within that time a tape/slide lasting longer than twenty minutes is likely to have undue influence. In any case, optimum concentration on tape/slides can only be maintained for about this period of time.

Tape/slides can, in fact, be modified by changing slides and additional recordings, or they can be used only in part. This is not totally satisfactory but is at least easier than attempting similar procedures with film.

This form of presentation can be more easily accepted and digested than written guides. The subject content can be the use of particularly large and extensive finding tools such as Referativnyi Zhurnal, Bulletin Signalétique or Chemical Abstracts. Alternatively it can be a refresher course on information retrieval in general or a first point of call for the reader who wishes to familiarize himself with a specific category of information or literature, such as statistics or government publications.

4.4.5 Films and Video Recordings

Films are less flexible than tape/slide presentations. It is unlikely that the teacher will find any of direct value in teaching information retrieval but he may find them useful as a reinforcement of lessons. A typical situation is a short course of one or two days' duration in which films offer an opportunity for reflection on information-retrieval skills between the more intensive sessions containing practical exercises or presenting ideas new to the audience.

Use of television can be recorded or live. One possible use is in the study of procedures and finding tools used to locate information. Thus the television camera and a suitably placed monitor can give everyone a near view of pages relevant to a case study. More realism may be possible if the camera is in the library rather than the classroom, with direction and instructions coming from the classroom. The advantages may not justify the difficulties of setting up and following through the exercise.

4.5 USE OF CASE STUDIES

The value of teaching information retrieval to users of scientific and technical information is based on simple theoretical arguments and supporting evidence from practice. Thus, one tends to study a number of practical cases, seeking common features and assessing the variable factors. Care is needed in choosing case studies. One case can be very different from another in many respects, and a single case cannot be representative.

Case studies of working situations in science and technology fall into five groups:

1. Successful projects
2. Failed projects (including disasters)
3. Users of libraries and information resources
4. Non-users of libraries and information resources
5. Successful scientists and technologists.

By implication there is a sixth category, the unsuccessful scientists and technologists, for which assumptions can be based on evidence from the other categories.

4.5.1 Successful Projects

For a case study to be really useful it should be in an area of interest relevant to the individual student. Often the most useful studies are those provided by information services, particularly when case studies provide illustrations for groups of mature students. At earlier levels of education, however, notable scientific achievements can provide more useful illustrations. These can include "commonplace" items, such as radio, television, watches, pocket calculators, vending machines, bicycles and pens, as well as the more obvious achievements in, for example, space research.

The primary material for studying the role of information in such achievements is not always easy to obtain. Many books and articles deal in generalities and personal roles, rather than in details. Also some details may have become obscured by time. One useful type of material results from serious study of particular innovations, which is often done by comparing successful and failed innovations. Innovation and development are complex subjects and differences between examples can be numerous because many factors besides information can play a role. This does not matter because the results of study provide useful background for groups of technology students.

4.5.2 Failed Projects (Including Disasters)

Some information on successful projects is difficult to obtain because publication may be thought to be of advantage to competitive manufacturers. By comparison, details on some failed projects may be difficult to obtain because organizations do not wish to publicize their failures. This is one reason why it is very difficult to estimate accurately the amount of unwitting duplication of research. Particular cases of unnecessary duplication and other failures are brought into publication through a number of sources.

One of the principal sources arises when the consequences have been very serious or disastrous, and there is a published report of an enquiry. This is usually detailed and gives factual information. The disadvantage is that the average scientist or technologist is more likely to be involved in personal failures, for example the surprise publication which pre-empts his own. Many scientists and technologists quickly publish notes or comment when subsequent publications unwittingly duplicate their own findings.

It is advisable to study in the first instance local reports of enquiries into disasters. In many cases the causes include a combination of human fault, information-retrieval fault and communication fault. One can then turn to the remaining published literature that is relevant to one's need.

A relatively simple case of an information-retrieval fault is given in Section 5.1 under 'Value of Information'.

4.5.3 Users of Libraries and Information Resources

A number of studies have been made of people who use libraries, the use made of them, and who does not use them. Professor Allen's studies in the United States are particularly notable in this respect. He refers to a certain class of users as 'gate-keepers' because they select and pass information to colleagues. One interesting aspect of such findings is the reason for this situation. It is possible that only certain people have developed adequate information-retrieval skills; but differences in human personalities also play a part.

Most scientists and technologists use libraries and information sources to some extent - in ways related to their own needs. Gate-keepers make much more use of them, often on behalf of colleagues. A third group uses these sources very rarely and is often dependent on gate-keepers and other intermediaries as sources of information.

It is sometimes worth studying the use made of the university or college library if records are available. Loan statistics can sometimes indicate patterns of use.

4.5.4 Non-Users of Libraries and Information Resources

People who do not use libraries often use informal information resources, primarily people. They find talking to people easier, pleasanter, quicker and more effective than searching the library's stock. A network of personal contacts has been referred to as the 'invisible college' and can be very effective, particularly for a group of researchers working in the same specialized field. A number of people with non-research interests also cultivate contacts, particularly when their output of work is to early deadlines. Many scientific writers come into this group. A professor of a postgraduate school in engineering once made the comment that the most important thing to teach students was how to establish lines of communication to people who could supply information. This is probably an overstatement, because the ability to use the information is more important and information supply could be made the responsibility of a gate-keeper.

4.5.5 Successful Scientists and Technologists

The achievements of scientists and technologists differ and the differences can be important. If common characteristics of the successful scientist and technologist can be identified, it may be possible to teach or instruct these as skills to degree students and practitioners. Some of the differences can be attributed to chance, but there are also differences in skill and ability. At a basic level one can consider two lathe operators or machinists. One of them may produce parts at a higher rate than the other, and often with a better finish. Although there may be a difference in manipulative skills, it is not simply the speed at which the controls are operated. Familiarity and knowledge of the next operation and the optimum setting of controls is significant. In a task which is unfamiliar to them, the operators have to relate the problems to experience, to exercise judgment and to seek advice. One of them may still prove more successful than the other.

So far as information-retrieval skills are concerned, one needs to decide the extent to which these are characteristics of the successful scientist or technologist. One can ascertain whether users of libraries and information centres include successful scientists and technologists. The other course of action is to examine the actions of successful practitioners. This can be done indirectly by reading biographies and autobiographies and noting relevant observations. Information-retrieval and library-use skills are often taken for granted or overlooked, just as the use of established laboratory equipment will be taken for granted.

4.6 EVALUATION AND FEEDBACK

Whilst much has been written on evaluation in the 'education' literature, little evaluation has actually been carried out in the field of 'user education'. Informal feedback and self analysis have been relied on mostly, often with the aid of a relatively simple questionnaire completed by the participants in a course.

Very detailed guidelines for evaluation have been issued by Unesco;[1]

The simple 'Participant evaluation report' shown in Figure 4.6 was used in field trials carried out whilst testing this Guide. In the 16 questionnaires completed in South Korea 26 per cent of respondents felt that their primary objective in registering for the course was well or mostly achieved and a further 69 per cent that it was mostly achieved. Seventy-five per cent of the respondents graded the course excellent or good.

Comparable courses in India and Indonesia seemed to be well appreciated, the major recommendations being that such courses should be developed at undergraduate and postgraduate level as integrated parts of university or other institutional programmes; and that, since in general participants found such courses quite a new experience, second-level or follow-up courses were very desirable.

Although such subjective judgements obtained so close to the course are of no quantitative value, they should not be dismissed too lightly and can certainly be considered in any post-course discussions between the course organizers, the host institution and course tutors. Without some feed-back, however inadequate, there is a danger of complacency.

Some feedback can also be obtained from any practical question sheets that are completed, although emphasis must be placed on the primary aims of the course, as given previously, in that a training programme exists to make the participants aware of information sources rather than to help them find the right answers to specific questions.

Perhaps there is truth in the theory that the best method of evaluation is whether people request, and ultimately attend, a higher-level course, or more important insist that their colleagues attend similar courses.

4.7 SELF-INSTRUCTION

Learning information-retrieval skills by self-instruction is not easy because of the numerous variables which make each case different. However, the teaching of information retrieval to users is a comparatively recent development and formerly there was little alternative to self-instruction.

Today, self-instruction is significant in two respects - self-dependence and self-service. In the former the user is dependent, perhaps temporarily, on his or her own skills, and in the latter he seeks to satisfy his own information needs by choice.

[1]"Guidelines for the evaluation of training courses, workshops and seminars in scientific and technical information and documentation". Prepared by F.W.Lancaster. Unesco, Paris, 1975. 63pp. (Doc.SC/75/WS/44)

UNESCO WORKSHOP

FINDING INFORMATION

Your candid reactions to the course will help us plan future courses. Your comments need not be signed, but PLEASE return the completed form.

1. Please state your primary objective in registering for this course.
    _____

2. Was this objective largely achieved through your participation in the course?

        Well Achieved   Mostly Achieved
        Moderately Achieved   Not Achieved

3. The topics which were MOST INTERESTING or most valuable for my purposes were:
    _____
    _____

4. The topics which were LEAST INTERESTING or least valuable for my purposes were:
    _____
    _____

5. What I liked most about the organization and procedure of the course was:
    _____

6. A suggested change or improvement in organization or procedure would be:
    _____

7. Do you have any suggestions for the improvement of the discussion process?
    _____

8. Did you find the practical sessions

        Too long: The right length: Too short:

9. Did you find the practical questions

        Easy: Moderately easy: Moderately difficult: Difficult:

10. What new topics might well be added when the next similar course is given?
    _____

11. What changes would you recommend in the physical facilities, travel arrangements, meals, lodgings, length of course, schedules, etc.?
    _____

12. In general, how do you rate this course?

        Excellent   Good   Fair   Poor

(Use the back of the page for any other comments.)

Figure 4.6 : Participant Evaluation Report

Self-dependence relates to availability of information advisers. It may be that instruction in information retrieval is given by short courses and that the teachers are not subsequently available for guidance. If other librarians and information scientists are regularly available, they can provide the necessary assistance and guidance.

The teacher with groups who will subsequently need to be self-dependent should provide adequate handouts and guidelines which are applicable to a wide range of situations (different needs and different libraries) for at least several years.

The two areas which should be covered by handouts are the procedures for searching for information, ideally listed as a number of steps; and choosing finding tools. In addition, such handouts should provide some indication of searching problems, particularly choosing subject headings. Of these three aspects - procedures, tools and problems - the second changes most rapidly with time. A two-level approach to it can be very appropriate. At the first level types or representative groups of finding tools are listed; at the second level specific items are given. Similarly, there can be two levels of instruction on search techniques - dealing with a general case at the first level and specific cases at the second.

There are other actions and emphases which the teacher can take when attempting to develop self-dependence. Rather than detail possibilities, it is better for the teacher to re-examine educational/instructional objectives and goals in relation to particular groups, and modify the goals accordingly. The next stage is to consider various methods of efficiently and effectively reaching those particular goals.

For the user who can consult librarians and information scientists a different situation exists. It may be better to concentrate on skills for satisfying 70 per cent to 80 per cent of information needs involving library use, leaving libraries and information centres to satisfy the remainder.

A simple search procedure is:

1. establish the search criteria;
2. decide what to do;
3. do the search in logical manner;
4. review the results and potential follow-up.

This is an adequate start and different groups will produce acceptable alternatives. The procedure can then be elaborated, eg. on the following lines for Step 1:

Establish the search criteria:

a. What is needed?
b. Is it clear?
c. How might others describe this information and related documentation?
d. How important is it? (Relates to available time, amounts of material and potential consequences of incomplete, inaccurate and unreliable information.)

Such an approach may appear to take an unwarranted amount of time and produce less than the ideal generalized search procedure. The apparent disadvantages can be fairly easily circumvented by using handouts for a different purpose. Supplying the handout subsequent to discussion can reveal oversights from the group's model search procedure, which can then be modified. It also shows that there is not a single ideal search procedure. Any teacher who lacks the conviction to teach in this manner should reflect on the necessity of preparing alternatives. Students should be taught how to compile their own lists and procedures. An information officer does not consult a generalized search procedure but in each search decides what to do and what to consult under what headings. Ideally, the normal user may be better following a simplified imitation of the same practice.

One difference between the information officer and the user lies in the range and depth of specialist knowledge about sources and location. Thus it is necessary to complement instruction in self-sufficiency with listings, particularly subject listings of the available finding tools. There is a problem that a proliferation of listings will cause confusion, and some people believe that the principle of a single starting point is still valid even if it is difficult to achieve. In the past, the library catalogue has fulfilled this role to a large extent. In the future, many selective lists may be prepared according to a wide variety of criteria, by sophisticated computer-based systems.

Where it is intended to develop self-instruction, it is necessary that library policy should reduce the causes of frustration to a minimum. Several elements of such a policy are implied in the provision of assistance or consultation, or of lists of finding tools and sources. On a more basic level it is necessary to ensure a suitable arrangement and availability of the material. Some finding tools are basic working material for some library operations, particularly acquisitions and cataloguing, and it is tempting to place these items for the convenience of librarians, particularly when they are the principal users.

So far self-instruction has been considered in general. Special cases include guidance in using finding tools, such as Beilstein, which are complex and need elaboration. In such a case a tape-slide or another detailed guide can be consulted when needed.

Part II Presentation and content

Chapters 5 to 9 are concerned with the presentation and content of courses. Two basic approaches are dealt with: a formal logical approach and an informal topic/project-orientated approach.

The first approach starts with the value of information and works through various aspects in a logical and sensible sequence. It is a formal and established approach suitable for both large and small groups (particularly large ones) and useful for students familiar with formal education. The three main chapters are as follows:

1. Fundamentals of Information Retrieval (Chapter 5)
2. Basic Methodology (Chapter 6)
3. Development of Search Techniques (Chapter 7).

Each of these sections can be allocated the same amount of course time - for example, one hour. It is possible to eliminate detail and illustrations and give the material in less time with supporting handouts. By contrast, it is possible to expand the programmes up to a total of thirty hours or more. Useful content material on particular topics has been provided by specialists in these areas and is attached as appendices. In addition, the supplementary notes in each chapter indicate various points to emphasise in courses.

In the alternative topic- or project-orientated approach individuals undertake a short project with a common theme or topic. The project is chosen to stimulate individuals to carry out certain tasks which involve information retrieval. The approach is informal, without a clear structure; it needs individuals who respond to the stimulus and groups of limited size. The project may extend into other aspects of communication skills and other tutors may then be needed.

The potential content of the two approaches is very similar. To avoid duplicating the content, the basic subject matter and the basic structure have been given in the sections dealing with the logical approach, this being the more established. The words "basic structure" or "basic outline" are referred to often in Chapters 5 to 9.

5. Course content 1—Fundamentals

The difficulty in describing possible material for instruction is the variety of circumstances that exist including, for example, the nature and size of the student group and the preferred objectives of the teacher. By making certain assumptions and indicating a general theme, together with alternatives and modifications, it is hoped to provide basic material for instruction in most situations.

The two main assumptions are a vocational viewpoint and an emphasis on the practical aspects of the instruction. In establishing a vocational viewpoint the teacher is concerned with teaching information retrieval as part of a vocation, ie. teaching civil-engineering information retrieval to civil engineers and so on.

This approach means referring to types of information need rather than types of literature, for example the description of one aspect of product information rather than trade literature in general. The practical approach is based on the belief that, for most people, information retrieval is of more interest as a practical skill than as a subject of academic study. It is probably more important for the civil engineer to know how to locate information than to give a discourse on the comparative merits of two or more bibliographical tools.

It is useful to work on a basis of a typical lecture or seminar of one hour, which makes ready for the practical work by explanation, guidance, and motivation.

A typical lecture structure for an initial session could be:

> General background
> - value of information
> - sources of information
>
> Recorded information (literature)
> - limiting factors and obstacles
> - keys to the use of a library
>
> Practical aspects

This could be the introductory lecture of a series, or could be confined to a single session. In the latter case, the section on practical aspects would become more important than the other two sections and would embody parts of other sessions described in later chapters. Otherwise the length of the three different sections would be determined by circumstances, though a useful theoretical basis is an equal period of time for each. Possible adjustments are fairly obvious: a highly motivated audience needs a minimal amount of time on the value of information and a well educated or experienced audience needs only a diagram for the sources of information.

5.1 TOPICS COVERED (LOGICAL APPROACH)

Using the suggested lesson structure, it is possible to give a commentary on the various topics for consideration - and also some illustrative matter, which can be used as lesson content.

5.1.1 General Background

A general background is necessary when the students have not received any earlier instruction or where such instruction has been of doubtful value. The topic has its greatest importance when students are not motivated or have an inadequate concept of the significance of information retrieval within their vocation. Conversely, the topic has lesser significance when motivation is high or the lectures on information retrieval have been preceded by lectures on closely related subjects such as communication.

The field of study has some effect on this session. For technologists, the value of information can be easily seen, and illustrated dramatically, with the help of case-studies; but for social scientists a logical argument may seem to be more appropriate. These variations can be more easily discussed after considering a general theme or presentation.

Value of Information

This topic has two important functions: one is to explain the significance of information retrieval to the students, the other is to put the students (and the teacher) at ease and in a receptive or activated mood. Some teachers may disagree with the former function, but most should be sympathetic to the second. When the course provides the first contact with a group of students or the first introduction to information retrieval for the students, use may be made of audio-visual aids such as tape/slide presentations and amusing illustrations. Experienced teachers will have their own techniques for capturing interest and will also know that in certain circumstances students can be resistant for such reasons as fatigue resulting from heavy lecture commitments.

A typical formal approach might be as follows:

- PREFACE

    - My name is ... The aim of this lecture series is to develop your information-retrieval skills in the field of ...
    - You should consider it as a practical subject, within this field, and treat it as you would other practical aspects of your subject.
    - In the lectures/seminar/ ... the reasons for learning information retrieval will, I hope, become clear to you and we shall be able to consider various basic applications and techniques.

- OUTLINE LECTURE (if not already given (e.g. on blackboard))

- VALUE OF INFORMATION

    - The theory is very simple - as the body needs food so the mind needs information. In simple cases the information is easily available, certainly if it can be remembered or observed.
    - Imagine for a moment that we prevented a person from remembering or observing information. In practical terms, loss of memory could be induced by drugs or loss of senses by covering eyes, ears, nose and limbs. The person would be seriously handicapped and the result can be anything from farcical to tragic. (Illustrate if necessary - humour if appropriate can be contrasted with tragedy in real situations when considering practical evidence.)
    - Thus a person unable to use observation or memory as a source of information suffers difficulty; one might examine whether similar handicaps result from being unable to use other sources of information.
    - In short, there should be practical evidence of people not being aware of published information and, conversely, of successes resulting from the ability to locate information in published literature.
    - Cases of both types exist.
    (Illustrate using appropriate examples. This is easiest in technology, using case studies and can be humorous or tragic as appropriate. The experienced and educated persons may prefer a dramatic case. For general situations use the following case:)

        'Chementator' (Chemical Engineering, Vol. 73, No.25, Dec 1966, p.69) reports that, "Failure to read the literature on titanium stress corrosion may have cost the space program more than $1.5M. The prime contractor for the Apollo manned space project used methanol to pressure-test twenty $100,000 fuel tanks for the Apollo spacecraft. Methanol's characteristics are similar to those of the fuels hydrazine and nitrogen tetroxide. After eighteen tanks had passed the 48-hour trial, the last two failed in test.

        "Hurried investigation after the fact .... showed that methanol causes stress corrosion in titanium, which was reported in the literature as long as ten years ago and again this April at a National Association of Corrosion Engineers meeting. An expert on titanium corrosion ... (reported) that as little as 2 per cent water in the methanol would have prevented the problem."

        "Neither the National Aeronautics and Space Administration, nor the ... (the contractor) nor the builder of the tanks ..., has decided what to do with the remaining fourteen items but they may be scrapped."

        "Now consider an attempt by an information officer to locate information on the stress corrosion of titanium by methanol. In less than 15 minutes an article by Shigetada Segawa and others can be located by using Chemical Abstracts (1965, 62:3767b). The article is 'The stress corrosion cracking of Ti and Zr in HCl-methanol solution'. Boshoku Gijitsu, vol. 13, 1964, pp. 214-217. The abstract was susceptible to stress-corrosion in the presence of methanol (and stress) and also indicated that an addition of 2 per cent water would solve the problem. The significance of this is that $1.5M would pay for a scientist/technologist/information scientist for substantially greater period than 15 minutes. At $15,000 p.a. the time period would be 100 years."

- Consider giving more details, using a case which can illustrate more obscure but important points: the relationship between research and information; the significance of hindsight; the occurrence and identification of information needs. An example is given in a study of U.S. military research projects called 'Project Hindsight'.

"In 1959, the United States Air Force introduced the C-130 transport aircraft into service. Five years later, in 1964, the C-141 transport was added to the fleet which by then had the C-130E variation. The C-141 exploited the potential of the available technology so that comparative payload and range of the three aircraft were as follows:

	C-130	C-130E	C-141
Comparative Payload	100%	140%	over 200%
Comparative Range	100%	160%	over 200%

The Hindsight studies calculated that each C-141 gave a saving of $8,926,000 over C-130 type. The total potential saving was $2.53 billion which is equal to more than 25 per cent of the total U.S. Department of Defense investment in scientific and technological research and development between 1948 and 1960."

- Link value of information to sources. For example, 'Information is in fact a valuable resources we ought to identify and compare the main sources and channels for acquiring it'.

Sources of Information

The important function of this topic is to put recorded information (literature) into the context of other sources of information. Some people attempt to seek information from literature when alternative sources may be better. Medical information needed by laymen is a classic example. When literature is unsuited for an information need, the result can be disillusionment. It is important to avoid such disillusionment. Attitudes to using literature and libraries often leave a lot to be desired in many areas. Because of these attitudes many people do not use libraries and literature as sources of information when it would be appropriate to do so. A partial solution to this problem is the provision of an effective information service. This can be expensive, but the benefits can more than justify the cost, even though other priorities may take whatever money is available. It is therefore necessary to provide guidance or permit discussion on the virtues and vices of the different sources of information.

The approach used might well be based on the following simple flowchart:

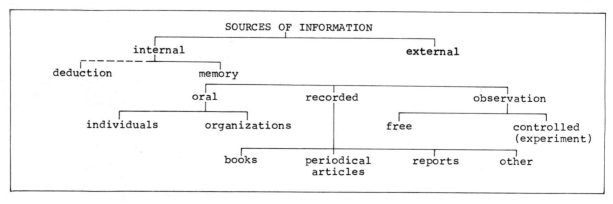

Figure 5.1.1 Sources of Information

For the newer or inexperienced student it is often desirable to build up the flow-diagram from discussion or a question-answer session. Some or all of the sources are likely to have been mentioned or implied earlier in the lecture/seminar, and the students may usefully learn by considering alternative choices themselves. If the sources have been implied, this response from the students can help the teacher modify the lecture beneficially. A typical lecture presentation may be based on the following notes:

- Establish the major sources of information.
- Having established these sources, the question arises as to which source one uses for a particular information need. This decision process is based on the fact that each of the sources has its own limiting factors and obstacles, with the result that the choice may or may not be obvious.
- Establish some of these limiting factors and obstacles, possibly by discussion;

  - memory; fast, portable, sometimes reliable (with well educated groups introduce the terms recall and relevance); but limited by experience (ask the questions; "How important is experience?" and "Why may it be important?") One can attempt to evaluate comparatively a number of variables for each of the main sources of information; but there are problems of measuring the variables. An easier alternative is to illustrate the differences by way of an example; consider all the telephone numbers of subscribers in the country's capital city. It is arguable whether a single individual can remember (i.e. recall) them all. Even if he can, the usefulness is limited, except perhaps as entertainment value, because the information is in a readily available source.
  - unrecorded specialist information. The specialist also needs information, for example to keep up-to-date. There can be obstacles in consulting authorities on certain specialized topics because they may be very remote.
  - recorded information; should be discussed in detail later but many people with knowledge of any value are often motivated to put it down on paper, where other people can refer to it at any time. (Can discuss the pressure to publish, with possible effects on quantity and quality of literature.)
  - observation; useful when the information is peculiar to local circumstance such as the height of a specific tree or the hardness of a particular piece of metal. It is likely that all information comes from this source originally, but observation involves investment of time and money. It makes sense not to repeat it without good reason. Hence the need to use the other sources of information to their full advantage.

- Consider the cases where the main source of information is not obvious, illustrating them where necessary:

  1. How does one boil an egg?
  2. What legal action can be taken against excessively noisy neighbours?
  3. What is the thermal conductivity of cod?
  4. What is Ohm's Law?
  5. When was the first artificial satellite launched?
  6. Why do rainbows occur?
  7. To what height do palm trees grow?

  Discuss the decision processes involved and consider the logic of one's memory, other people's memories, society's memories (libraries) and the original source - observation. Also point out that some sources of information may work in combinations.

- Perhaps the most under-utilized major source of information is recorded information. There is a second decision process: does one use books, periodical articles, reports or other records? One can discuss the relative virtues of each as a source of information. This involves the differences between librarianship, which is concerned with different forms of material, and information retrieval, which is concerned with different information needs. The currently unavoidable difficulties in matching information needs and forms of material leads to some unfortunate circumstances. If time permits there can be an interesting discussion on the use and usefulness of conference proceedings, review articles, and other items which sometimes create anomalies. This topic can lead usefully on to the next section which considers recorded information in more detail.

One problem in discussing sources of information is terminology. Precise terms are not likely to be familiar to the student, and the teacher must decide what is acceptable and clarify any vagueness. A second problem is the danger of digressing or going into too much detail. Indeed it would be unusual to discuss all that has been included in the foregoing outline, which lists items worth considering for inclusion. One should aim for a balanced presentation appropriate to the situation.

The terms used in Figure 5.1.1 are worth further elucidation. Firstly, the term 'deduction'. An acceptable synonym for it is 'reasoning', but 'guesswork' may be a different matter. It is suggested here that guesswork is a specific aspect of deduction and may be acceptable as a source of information. If used as such, its reliability and accuracy are possibly low and for specific cases the student ought to be able to decide whether or not it is acceptable. Its relevance to information retrieval is that it may sometimes be a lazy but less satisfactory alternative to recorded information. This danger is probably greater during employment, particularly under pressure of work.

The topic of deduction is on the fringe of information retrieval and is advisedly linked to the major sources of information by a broken line. Additional material on it may be supplied when dealing with particular subject groups, notably mathematics. More experienced students may express interest particularly in seminar or discussion sessions.

Secondly, the term 'memory'. This has synonyms and near-synonyms in experience, knowledge, and know-how. Some students may place a high value on experience and it is often a useful practice to develop a question-and-answer session by asking if experience is important and subsequently why it is important. They are more likely to understand the circumstances than to note (and probably forget) the source of information. The cost of this learning is the time taken to work through the sources of information and the teacher must think carefully in the initial planning.

The term 'oral sources of information' indicates the means of access to the contents of other people's and organizations' information resources (memories and files). 'Oral sources' can be suggested through people, colleagues, friends, acquaintances, personal contacts, specialists and authorities. Even using the terms "oral" and 'recorded' information, leaves an intermediate category for telephone messages and divides live radio and television transmissions from recordings. It must be left to each teacher to use the terms he thinks most appropriate. A number of alternative sub-groups of oral sources are possible. The use of individuals and organization (or corporate bodies) is suggested because many libraries and published directories make a distinction between them.

The term 'recorded information' covers not only literature and libraries but also the increasingly available data banks and non-print media. Terms such as libraries, literature, are no longer sufficiently comprehensive. Difficulty may be experienced with sub-classes of recorded information. Once again the teacher should use the terms he believes to be most suitable. Books and periodical articles are likely to be agreed terms but when teaching information retrieval it may be useful to question even these. So far as library handling is concerned the units are books and periodicals. Periodicals are too broad a category for the bulk of the specific information needs of most library users. It is better to consider each periodical article as a separate unit. Some special libraries follow this practice and index each article, as do the publishers of abstracting journals.

The same argument may not be true for books, and some thought must be given to the ideal unit of information retrieval for this material. In the case of reference books in particular the 'chapter' or 'section' is probably the unit which has to be indexed to the required depth. A more distinctive category may be conference proceedings, where individual papers are the items specifically needed by most users. The criteria for deciding the sub-classes of recorded information are inevitably influenced by their treatment in compiling bibliographical tools such as abstracting and indexing publications.

At least one other specific category is needed - for publications which are not necessarily treated as books or periodicals, notably reports, which are important to scientists and technologists.

The final sub-class, 'other', is a little unfortunate as it suggests that one cannot satisfactorily define the various constituents. In practice there are a lot of possible forms for recorded information and it would be time-consuming to consider them in detail, particularly as each group may have its own bibliographical tool to locate specific items. The person using these other forms of recorded information is usually aware of those that he may need, for example, news items and maps. One exercise which can sometimes be carried out by teachers or students is to list as many possibilities as one can in sixty seconds. There may be little instructive value in such an exercise but it can be useful for 'interest lectures' or seminars, such as may be given to school children or teenagers.

The term 'observation' can cover free observation or controlled observation (experiment). It is intended to be widely interpreted and includes information received indirectly through instruments as well as through the senses. An interesting group discussion can be developed, and can lead to useful learning, provided that the teacher is concerned only with aspects of information and communication; otherwise this would be a digression.

5.1.2 Limiting Factors and Obstacles in Using Recorded Sources of Information

Each of the major sources of information has its own limiting factors and obstacles. If this was not so, one major source would be sufficient.

The purpose of describing the limiting factors and obstacles to using recorded sources of information is to make the students aware of them and able to relate their effects to the likelihood of satisfying his information needs. Different teachers may have different ideas of what these limiting factors and obstacles are. This does not really matter; what does matter is that the students subsequently become more effective because they have acquired beneficial skills.

A useful technique in presenting these limiting factors is to present them in precise form in a handout and to illustrate them within the lecture or seminar. The handout can contain the following information:

LIMITING FACTORS AND OBSTACLES

Limiting factors   - ability to define the information needs
                     - individual skill in information-retrieval methods
                     - knowledge of the subject and its terminology
                     - knowledge of bibliographical tools in the subject
                     - time needed for searching
Obstacles   - bulk (literature 'explosion')
                     - scatter - by source
                                - by language
                                - by topic within source
                                - by physical location
                     - variation in descriptive terms
                                - differences between concepts and words
                                - differences between specific and general considerations
                                - synonyms

This list, whether in handout form or not, is little more than a statement, and some explanation for students is necessary. The following paragraphs attempt this.

Ability to Define Information Needs. If one does not know what one is looking for, there is only a low probability of finding it. One is always working on probabilities in seeking information and, as a general rule, the probability of success increases with ability to specify need - and vice-versa.

Individual Skill in Information-Retrieval Methods. This is the specific skill of finding a suitable approach and strategy. An analogy is useful here: it could be in wood-working, where a combination of tools is used along with skill in choosing and manipulating them to achieve the objective.

If the location of information involves particular skills, the argument may be presented that such a task should be for a specialist, that is, the librarian or information officer. This is sensible in a number of cases, particularly where the specialist's knowledge of, or proximity to, the material is the key factor. However, in other cases, the key factor may be subject knowledge. In others still, information specialists may not be accessible. Users are therefore well advised to learn some rudimentary information skills and sources.

Knowledge of the Subject and its Terminology. In addition to information-retrieval skills there is often a need to know the meaning of words associated with the subject. Words denoting author and subject are used by indexers as 'tags' to enable relevant material to be brought to the notice of the information seekers. Language is such that searching - matching words with index terms - is sometimes difficult. Also, the subject of an article is unlikely to match exactly the needs of the searcher. So relevance is often a matter of degree, and subject knowledge may be necessary to judge that degree.

Knowledge of Bibliographical Tools. One may also add 'familiarity with bibliographical tools'. The term bibliographical tools is a precise one but it may be better to use the term finding tools (as in this Guide) which is indicative of function even if it is not so precise.

Time Needed for Searching. It takes time to seek information, especially when uncertainties exist. In these circumstances one progresses by working on probabilities: the probable risks, benefits and likelihood of productive results. The time spent on searching should be related to these probabilities, and it can be considered as an investment.

Obstacles - Bulk and Scatter. These can be well illustrated by comparing the stocks of a number of libraries in terms of the amount of time taken to search the complete collection.

The bulk problem is illustrated by the time taken to search all the stock of a library which is very large yet does not include all the available material. Reference should be made in the lesson to estimates that literature output is doubling every 12.1/2 to 15 years, and a distinction should be made between literature and information. In practice it is the **exponential** growth of literature which is the problem rather than any information explosion. The teacher can indicate how absurd it is that so many people see publication as the objective of much of their endeavour but pay scant heed to using material published elsewhere.

Time Taken to Scan Titles Held by Libraries of Various Sizes

	Personal Bookshelf	Small University Library	National Library	Largest Library in the World
Books	4 mins.	4 weeks	4 years	40 years
Periodical articles	40 mins.	4 years	Up to 400 years	400 years
Reports	4 secs.	4 months		

Sequential Search Rate: 400 titles per hour

N.B. These values are only rough approximations intended to show the orders of magnitude

Figure 5.1.2a : Table for Use in Illustrating Bulk and Scatter of Literature

Figure 5.1.2a illustrates part of the 'scatter' problem. Scatter by source refers to the fact that material in this context can include books, periodicals, reports, etc. or indeed a much wider range of material. It should always be possible to identify source, which can include the organization producing the information. Figure 5.1.2a also illustrates scatter by physical location.

Language scatter is self-explanatory and in itself often imposes a definite constraint on the enquirer, depending on his linguistic capabilities.

When considering the scatter of certain topics within sources it is necessary to make brief reference to classification. Material is often located according to the topic sought, but certain topics can exist as part of a variety of subjects. For example, instrumentation may exist as a subject on its own and also be scattered in aeronautical engineering, automobile engineering and so on.

Variation in Descriptive Terms. This difficulty arises through limitations of language, particularly the problem of specifying concepts. In practice, concepts and words do not necessarily match. For example, an information officer was asked through an intermediary to provide information on models of chemical plant. After information was located on mathematical models the enquirer explained that he actually wanted a cardboard model. It is very easy to overlook this important obstacle because in most communication fields it is a lesser problem: single words rarely become 'key-words' and the meaning is usually context. It is possible to illustrate this point with examples which are very easy to find. A 'Stirling heat engine' may also be described as an 'air engine' or a 'heat engine'.

These examples are real enough but they may not make the impact needed. Amusing illustrations may be more memorable. Figure 5.1.2b shows how the phrase 'perfect aeroplane' will produce a number of different concepts according to the background of the person.

Alternative examples can be found in machine translation - a subject with close links with information retrieval. Anecdotes occasionally spread through the information field about the latest 'mis-match.'

A common experiment in machine translation is to feed English language material into a computer to produce a Russian translation. This process is then reversed to give a final output in English. The initial input and the final output are compared as an effective measure of the system. The phrase "the spirit is willing but the flesh is weak" came back as "the vodka is good but the meat has gone off". In another case "out of sight, out of mind" came back as, "invisible idiot".

5.1.3 Keys to the Use of a Library

Recorded information is available in large amounts and in many forms. To permit effective use of the material substantial portions are acquired selectively and made available in libraries. The libraries, publishers and other then attempt to provide 'keys' to the use of this material.

The three main keys considered here are techniques for reducing the amount of unnecessary browsing and for providing listings which indicate the location of relevant

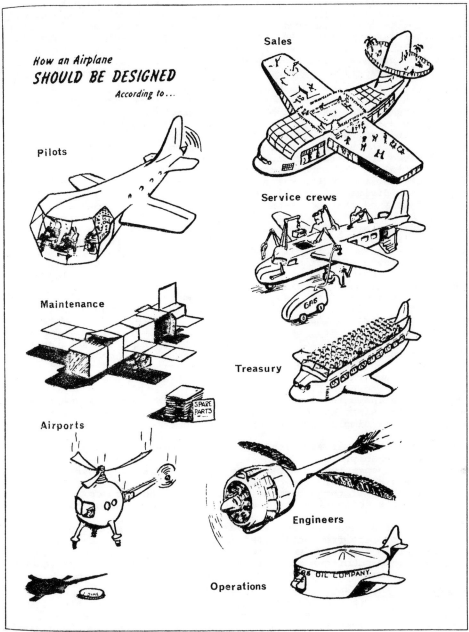

1 Desirable features of a modern aircraft—liberally interpreted

From: HOWARTH, F. The aircraft industry. Metals and Materials, vol.2, no.4,
April 1968, pp 110-119. Reproduced by kind permission of the author
who has now used the illustration for some 20 years but does not have
knowledge of its true originator.

Figure 5.1.2b   Desirable Features of a Modern Aircraft-Liberally Interpreted

material. One technique is to group together books of a similar content, another is to list in logical order the authors and subjects of the contents of a collection, and the third is to provide a wide range of listings and associated guides to the world's literature and other forms of recorded information. It is as well to elaborate on this third aspect, and the teacher can discuss the other two techniques as he or she sees fit. Notes of these three techniques are presented here for guidance under the respective headings of classification, cataloguing and indexing, and bibliographical finding tools.

Classification. In a general collection of material, various forms are available, but books are usually the largest part. It has already been shown that searching more than a few shelves of books sequentially is time-consuming. It is desirable therefore that books on the same subject are put together and adjacent to similar material. It would be possible to put all the scientific books in one room, say room 5, one corner of which would be devoted to general science, another to mathematics, a third to physics and so on. The material in the corners could be further sub-divided into topics, so one table designated table 2 in the mathematics corner could contain books on algebra. Of course further sub-division is still possible. A reader then wanting material on algebra would need to know that books on algebra are in room 5, corner 1, table 2. This information could be provided by a simple list near the entrance hall.

The numbers could be grouped together as 512 once the system is explained. What we are now discussing is known as classification and this oversimplified outline could be used by many stores - including retail shops - for ease of access. Classification is also used for many aspects of sorting knowledge, components, and items. Examples in science include the periodic table of elements and the taxonomy of biological forms. Similarly many individuals and organizations have devised simple classification schemes for making available components such as screws and bolts without needless searching.

When considering the classification of books in libraries, there are some practical elements which make changes to the earlier outline desirable. If the system is followed rigidly, then some rooms, corners, tables, etc., will be overcrowded and others inefficiently filled. Indeed if the standardized schemes which exist are adopted then there may be insufficient rooms, corners and tables. So long as the number codings represent subject classification numbers and shelf numbers there is no need to adhere to actual geographical position and this is not normally done. Likewise the length of numbers need not be restricted to three significant numbers.

There are other refinements. Some can be mentioned in lessons as being significant for the reader and others may be pointed out to certain groups. Many of these additional refinements are made for simple practical reasons and are best shown on a tour of the library.

All the variations of classification cannot be covered, and they should not be, for that is really librarianship.

Knowledge does not conveniently break down into a number of separate and mutually exclusive sections and subjects. Some classification schemes make use of letters and this permits an extension of the number of sections. In other cases letters can be used as a prefix for separate collections to the main collection. Colour coding can even be used.

In some classification schemes, certain number combinations have particular significance. Such combinations are frequently used to indicate geographical data.

For the reader it is necessary to appreciate that classification schemes have advantages, particularly in producing a logical sequence, but there are limitations. It is important to know that one section or class is not mutually exclusive.

Cataloguing and Indexing. Classification has its limitations, being insufficiently precise for locating a specific book or other item. It would seem sensible to supplement classification with other techniques. One common technique is to 'tag' each item with words which can be useful identifiers. One obvious group of tags consists of the names of the authors and others responsible for producing the various publications. If the tags are filed or listed in a logical order, they help create a useful finding tool or, more specifically, a bibliographical tool.

In addition to author tags, subject word tags (index terms, descriptions, identifiers, keywords) are commonly used. Although subject terms do not produce logical arrangements they can be precise in many cases and are therefore complementary to classification.

Other facets can also be used as tags, for example the titles of books, periodicals and report series. In the case of books the title can be used as a substitute for subject indexing. In effect, publishers' catalogues serve as bibliographical tools in which the name of the publisher is used as the primary tag. The whole field of cataloguing and indexing is a complete subject within librarianship and information science. The library users' main interest is the subject approach and at least one publication deals with this approach.[1] This includes limitations, advantages and typical schemes.

Bibliographical Tools. Catalogues are found in most libraries and serve as a finding tool for their book collection. A single book collection is only part of the total available recorded information. That total is too large to index or catalogue in one sequence for all the needs of the world; some of the various reasons have been given already. Around the year 1900 two men started to index the world's literature. They never progressed beyond establishing a classification scheme, now known as the Universal Decimal Classification (U.D.C.).

The current situation can be illustrated by a simple diagram (Figure 5.1.3a). Bearing in mind the fact that the quantity of literature is far too large to search through completely for relevant information, classification helps by bringing together associated material and permits browsing to be fairly efficient. Cataloguing a library's book collection gives greater precision, but as a finding tool is limited to a small portion of the totally available literature. Other finding tools exist which cover other collections and other subsets of the total literature.

Together with reviews and the compilations in reference books, these finding tools constitute what is known as secondary literature. The finding tools vary widely in size, coverage, scope, presentation, and quality. The scientist or technologist is likely to become familiar with at least some of the finding tools in his own subject areas, and this familiarity is important because it makes for quicker and more effective use. However, it is insufficient in itself, since individuals frequently have to search outside their own subject fields or beyond the finding tools that are known to them. Any list of finding tools given here, would be incomplete, soon out of date and wasteful of students' time and teachers' effort. Such lists and guides do exist and are published. Thus we have a tertiary literature or 'guides to guides'. In spite of their usefulness, one needs to be aware of their limitations, including variations in quality. New publications, including secondary literature, soon make tertiary literature out of date. Regular publishing offsets this to some extent; but publishing delays build up so that it is never possible to be fully up to date.

At this stage, it is useful to show students what these 'guides to guides' actually look like. Examples of two basic types should be shown: one being subject-orientated and the other being an indexed list of bibliographies, reference books, or abstracting journals. The subject-orientated guide can usefully be in a subject field of common interest to the students of the particular group.

5.2 MODIFICATIONS AND CONSIDERATIONS ARISING FROM DIFFERENT CIRCUMSTANCES

5.2.1 Practical Aspects

The content of this part of the course will depend very much on circumstances. If there is only a single lecture, practical aspects will form a substantial portion of it and will contain much of the material from Chapter 7. If the single lecture is descriptive or for general interest, a prepared audio-visual presentation is useful. If the single lecture is intended to be instructive - which is difficult under these limitations - the whole of it can be on practical aspects, with substantial support from handouts.

If the session is to be followed by practical work, it is possibly best to make it a briefing session. For undergraduates, this briefing will be directly related to the practical exercises. For postgraduates, or comparable groups, the practical sessions need not be so direct and should be more concerned with aims and objectives.

Some countries may possess the ideal combination of a rich literature and modern library systems; others may be subject to financial, geographical, historical, and language problems. These differences may need to be taken into account.

5.2.2 Subject Differences

Some course groups may not have a common interest. It can be argued that this is an advantage because the interaction of different subject specialists can extend the range of ideas and give greater stimulus to the groups.

For university and college courses, a common subject interest is usual and this has yet to be shown as a real disadvantage. However, the teacher should take the common subject interest into consideration in order to establish a sympathetic understanding with the group more easily.

---

[1]Foskett, A.C. Subject Approach to Information. 2nd edition. London, Bingley. 1971. 429p

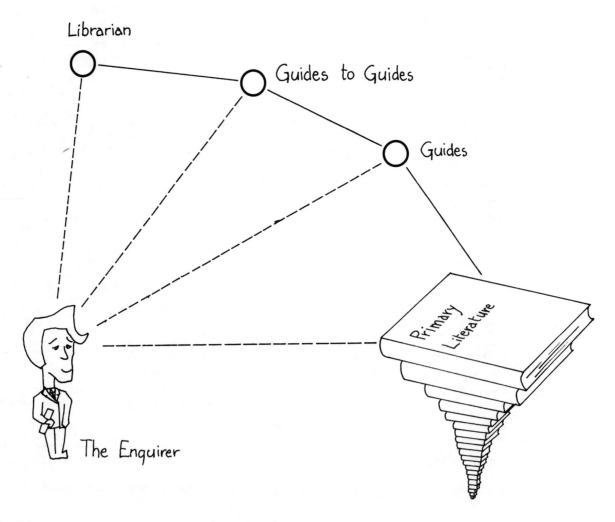

Figure 5.1.3a   Relationship between enquirer, librarian and the literature

LIBRARIAN	TERTIARY LITERATURE (GUIDES TO GUIDES)	SECONDARY LITERATURE (GUIDES)	PRIMARY LITERATURE
Selection	Guides to the Literature	Library Catalogues	Books
Acquisition			
Classifying	Bibliographies of Bibliographies	Book Lists	Periodical Articles
Indexing		Publishers' Catalogues	
Making available the literature	Guides to Abstracting Services	National Bibliographies	Reports
Instruction		Bibliographies	Pamphlets
Guidance		Review Articles	Literature with a specific function:
Information			Standards
		Abstracting and Indexing Services	Patents
			Newspaper Items
			Maps and Atlases
			Timetables
			Telephone Directories
			Theses

Figure 5.1.3b Using the Various Sources of Information

51

While it is not possible to give comprehensive notes on the significant variations for each subject, the following notes draw attention to some key aspects which, together with the advice and comments of course tutors and possibly subject specialists, should be sufficient. The order of scientific and technological subjects is approximately that used in the Dewey Decimal Classification Scheme.

General Science. Information retrieval is important in avoiding unintentional duplication of research and in advancing the field of knowledge. Duplication of research is a waste of time and money. Advancement of the field of knowledge would seem to depend on familiarity with the existing knowledge in a particular area, but the view is sometimes taken that existing knowledge can restrict creativity. Certainly, care needs to be taken that earlier work does not constrain one's imagination; but, conversely, reading widely can itself be a stimulus, for ideas can originate in apparently unrelated fields.

Most scientists recognize the need for information retrieval, but the tendency has been to depend on oral sources of information. The literature provides an equally important formal source of information, and it has been relatively neglected by many scientists because they have not received the necessary instruction. One phenomenon is the scientist who is able to acquire information from literature and often acts as an intermediary or 'gatekeeper' between literature and other members of a group or even 'an invisible college' (see Section 4.5.3).

Mathematics and Computer Science. Although some people may prefer these two fields to be considered separately, it is convenient to group them here because of the common interest in manipulating relationships. Both fields (like information science) benefit society by their application to a wide range of topics. These applications can result in equally wide-ranging information needs.

The basic course structure can be modified to permit discussion of the relative merits of reasoning and literature as sources of information in more detail because reasoning (or deduction) plays such a key part in these fields of study. In contrast, attention can be drawn to the wide-ranging application of mathematics and its related information needs. Finally, the role of probability in information retrieval may be appreciated by mathematicians. At certain educational levels reference can be made to 'precision' and 'recall'.

Astronomy. Observation and certain kinds of atlas and handbook are important in routine astronomical work. Examples of these reference books may be useful in illustrating points in the lecture. Entertaining and illuminating examples can be found in science fiction and are suitable - up to a certain educational level - for motivating the audience.

Physics. In this context, what applies to science in general applies to physics.

Chemistry. The literature and bibliographical tools of chemistry are notably well organized and comprehensive. In addition, much of chemical knowledge lends itself to classification and chemists generally accept the literature as important. The result is that teaching information retrieval to chemists appears to be straightforward, but there can be problems. Some chemists may be optimistic about their existing knowledge of information retrieval. Practising chemists may put greater emphasis than others on subject knowledge for literature searching. Finally, a group which readily accepts the importance of information retrieval can demand a greater depth of course study.

Geology. Although there are substantial practical differences between geology and science in general, these need not greatly affect the basic course structure. One possibility for discussion, particularly with a questioning audience, is the currency of geological information. This is one of the fields where the literature retains its value for a relatively long period, certainly much longer than electronics and aeronautics.

Life Sciences. Like chemistry, some parts of life sciences lend themselves to classification, and have good bibliographical tools, while scientists in this area soon accept information retrieval as an important aspect of their work. Like geology, the literature has a substantial period of currency.

Engineering and Technology in General. Technology makes use of scientific knowledge and innovations in various fields to improve and maintain the health and material wealth of society. Thus, the literature serves technologists as a reservoir of knowledge. In turn, society uses the literature of technology when studying its welfare needs and the implications of technological developments. Information retrieval is really a part of the communication process, and this can be seen most clearly in technology and its relationships with the physical and social sciences. Anyone who doubts this should read some reports of enquiries into the causes of disasters.

Materials Science and Technology. Classification schemes used locally should be checked for their ability to classify materials science. Commonly used classification schemes scatter books of interest in the fields of chemistry, physics, and engineering indicating the origins of materials science. Other common schemes are simply antiquated. The result is that the teacher should ensure adequate mention of the scatter problem and the limitations of classification schemes.

Mechanical, Civil, and Production Engineering. In this context, what applies to engineering in general also applies to these fields, and the basic structure of the session is ideally suited to them. Case studies of innovations and disasters are relatively easy to find and are suitable for the 'value of information' topic.

For these fields account should be taken of the differences between the different types of work performed by engineers, such as design and maintenance. Another feature is the difficulty of access to certain types of information. Process and material costs are classical examples, and further improvements are needed in product information. These are essentially practical aspects and are discussed in greater detail in Chapters 7 and 8.

Aeronautical Engineering. In this fast moving field, financial investment is high. On the technological side the features are similar to those of other engineering fields.

Electrical and Electronic Engineering. Electrical engineering is not very different from other engineering fields except that one particular aspect has grown into a rapidly developing field in its own right. This field is electronic engineering: it has wide-ranging effects in engineering and has tended to influence electrical engineering as a whole. Its rapid developments have led to literature, particularly of periodical articles and reports, which has a relatively short period of currency. Keeping up-to-date should be given prominence.

Chemical Engineering. The general aspects of engineering apply to chemical engineering but two additional points are worth noting. One is the need to exploit the well organized literature of chemistry and the other is the need for information from other fields such as materials technology, computer science, and management.

Medicine. Information-retrieval problems are fairly standardized in this field although there is a significant difference of approach between research and professional medicine, and this should be indicated with an emphasis on confidentiality when dealing with the latter. The terminology is well established and documented, but there are problems in the use and definition of proprietary names of drugs.

Agriculture, Horticulture, Forestry, and Land Management. Practitioners of the various forms of land management may be remote from adequate libraries and information sources, but they still have important information needs. The best way of satisfying such needs may well be to place intermediaries between the users and the sources. Such people can be advisers, interpreting both users' needs and available information. Circumstances vary and the teacher needs to examine the local situation and modify his instruction accordingly. Within the basic structure of the session, the question of 'who searches for information?' can be discussed in more detail.

Food Science and Technology. A group with a common subject interest as specific as food science and technology needs to ensure it is not being constrained by taking too narrow a view. It is advisable, with highly specialized groups, to stress the significance of general/specific relationships at this early stage. The different levels of specificity and different levels of literature are important and time should be allowed for groups to accept and gain familiarity with them when discussing or carrying out practical exercises.

Management. It has been suggested that information retrieval is part of management and, if this is interpreted as an 'enabling function', this is true, especially if 'self-management' is included. Management groups need extensive material, including material from 'management information systems'. Other differences will depend very much on particular circumstances, and the teacher is advised to consider his objectives carefully with such groups and modify the basic structure of the session accordingly. The basic structure is probably more suited to students rather than practising managers.

Social Sciences, Humanities, and Arts in General. In some cases the basic structure of the session may be adequate with little change; in others a completely different approach may be more suitable.

Social Sciences. A social scientist ought to accept that the library is his workshop and that less emphasis can be placed on the value and sources of information than with natural sciences or technology. More time could be usefully spent on the extensive information and literature needs of social scientists. One social scientist said that everybody's rubbish was their working material.

Economics. The basic outline is reasonably satisfactory, probably more so than for other non-scientific or technological fields. The study of successful and failed economic policies, in order to assess the role of information and its communication, could be interesting.

Humanities. There is a tendency for information-retrieval instruction to be given first to scientists and technologists and then extended to social scientists. One of the neglected areas can be the humanities; yet their contribution to society is important and should be on a sound basis implying a thorough study of the background. Much needs to be done.

Arts (including language and literature). The view has been given that arts students acquire a useful working knowledge of libraries through their studies and the guidance of tutors. Without a proper understanding of the use of literature, the students are likely to produce work which lacks originality or substance. This Guide is based on the thesis (and experience) that instruction is necessary for the proper understanding of the use of literature. The different needs and uses by arts students may suggest a different teaching approach.

5.2.3 Differences Arising from Educational Level

The main groups we are concerned with are:

- Undergraduate
- Postgraduate
- In-service or further education.

Undergraduate Level. The basic outline for the session is orientated at this level.

Postgraduate Level. Compared with undergraduates, postgraduates tend to be more experienced and smaller in number, as well as older. They are thus able to cover the same ground more quickly, in greater depth, or both.

As a standard pattern the following modified form of the first lecture has been found suitable:

General Background - Value of Information
                      (use studies made of information in their field)

                    - Sources of Information
                      (present diagram - Figure 5.1.1;
                      give reasons for flowchart, choice of source)

Recorded Information - Limiting Factors and Obstacles
                      (brief listing - look for feedback)

                    - Keys to Use of Library
                      (brief description to give logic of search tools)

Practical Aspects   - (case/study and briefing for first practical
                      session - see notes in Chapter 7).

The result of this particular approach is to reduce to a minimum the sections on sources of information and the limiting factors and obstacles of recorded information. This allows more time to be spent on practical aspects.

In-Service or Further Education. This level is similar in some respects to the postgraduate level. The chief difference is that some in-service instruction is to people who are not immediately receptive to the approach in the basic outline. If this is the situation, an alternative approach must be found. One which has proved satisfactory in our experience starts with an attractive tape/slide presentation based on a case study and follows with a brief discussion of information:

- What is it?
- Has it any value?
- Where do we get it from?
- What are the problems?

This should take about forty minutes, half of which should be given to the case study. The important features are attractive presentation, brief statements of the situation and emphasis on practical aspects.

5.2.4 Differences in Teaching Approach

Formal Approach. The basic structure lends itself to a formal approach - a lecture to a large audience. In fact, the audience can be very large and of a mixed subject interest. If a single session on fundamentals can be arranged, more teaching time may be available for direct practical instruction. If a single session is spent on fundamentals, the teacher must not undervalue the fundamentals or overvalue his own lecturing ability. The prime objective in the fundamentals is to attract the attention of the group, create interest, and develop the idea of systematically using information as a valuable resource. Adequate preparation is necessary and it is desirable to use visual material to illustrate key points.

The formal approach need not be interpreted as a traditional lecture by a single person for a single hour. A more imaginative person may try different variations - two or more lecturers, audience participation, and visual demonstrations - to achieve his prime objective. However, novel approaches can go wrong and are better tested on smaller groups where the consequences can be less embarrassing and less disruptive to the prime objective.

Informal Approach. The basic structure can also be used with an informal approach. When to use the informal approach is best determined by the make-up of the audience and by the teacher's own personality. Size of audience is a key feature, and an informal approach is likely to be used for groups of less than twenty students.

The lecture outline can be much the same as with the formal approach but may include contributions from the students. Timing and details are adjusted to the situation, rather than planned, and a more casual approach is adopted. The differences between the two approaches may not be very great, but it is useful if the informal approach can be adopted when the situation merits it, even if it is not specifically planned to be so.

Practical Approach. The basic instruction in information retrieval must be a balanced combination of practical exercises and theoretical exposition. Practical aspects need to be included because an essentially practical skill is being developed. The balance of theory and practice may change in different circumstances. The basic instruction is already biased towards the practical content without jeopardizing understanding of the theory.

The fundamentals can be taught in a more practical manner than by the basic structure. For example, students may themselves examine case histories, use question-and-answer sessions for sources of information, or make calculations about the quantity of literature and the probability of locating an item. However such a practical approach uses more time and can mean going into greater depth than is normal in lectures. It is therefore most suitable for a limited number of postgraduate students receiving in-depth instruction. Such an approach is described in detail in Chapter 9. Alternatively, a teacher may start the instruction by simply describing bibliographical tools and search techniques, and then lead into practical applications. However, this is not advisable unless the teacher is absolutely certain that the students understand why they are being taught the subject, i.e. that the bibliographical tools being shown are adequate for their needs and that they can adapt their skills, once learned, to satisfy information needs not anticipated in the instruction.

Theoretical Approach. Some theoretical content is necessary for such an understanding. In the basic structure the balance of theory and practice has evolved from experience, and is believed to be about right. If the basic structure is used, it is advised that at least one third, and ideally one half, of the available time is spent on practical aspects and that at least the key points of theory are given. Where time is minimal, some of the theory and details of more specialized practical search techniques can be given in a handout.

Occasions may warrant an extension of the theoretical approach, eg. where a group of students require instruction in greater depth than in the basic structure and sufficient time is available. Alternatively, a group may simply wish to explore different approaches, or may be so large that an extension of the theoretical approach is necessary.

In a theoretical approach a group can study such topics as:

- the nature, occurrence and identification of information needs;
- decision points and processes;
- effects of variables.

The nature, occurrence, and identification of information needs have not yet been fully defined, but it is possible to give some indication of the circumstances in which needs arise. The principal circumstances are when a novelty element is introduced into some sphere of work. The word 'nova' is used to describe this novelty element. 'Novae' can be subdivided into categories: a new job; a new field of study; a departure from traditional or planned practice; a new product, process, or idea which relates to existing work. (See Section 5.3.1 for details.)

Not all information needs result from 'novae'. The human memory is limited and has preferences for certain forms of information. Even routine work makes use of recorded data, statistics and other facts which fulfil information needs. Once a need has been identified, one or more decisions have to be made. Should the information be sought? What major sources (oral, recorded, observed) should be used? - and so on. Attempts can be made to use flow-charts and similar aids to decision-making.

Information Science Approach. When teaching students of science and technology, one can approach information by using the literature of their own fields with appropriate practical examples. Instruction can be limited to aspects useful to the vocations of the audience, in the same way as an engineer is taught familiarity with the properties of materials so that he can use them to the best advantage in his work.

The differing views of the nature of information have been clearly described by Anderla[1]. The traditional approach regards information as both an input and an output of research. A broader or socio-cultural view regards information and transferable knowledge as the same. Thus information transfer is defined as the transfer of knowledge serving a wide range of activities. It is rapidly gaining acceptance that information is a resource, as fundamental as energy or matter, which affects all human activity and acts as an indispensible and irreplaceable link between intellectual and material activities. This view is very acceptable and, therefore, attractive to use in teaching information retrieval to people of various vocations.

If students accept this view then the significance and value of information follow very easily. This approach can be called the information-science approach, but many problems may arise because of the difficulty of measuring information in terms of cost, value, and even quantity.

Adopting an information-science approach can influence the introduction to information in other ways. The structure of literature can be taught in a different manner and the secondary and tertiary literature can be described in more detail. Experience has shown that faults in the tertiary literature can create some difficulties; the main faults being variations in quality, delays in publication, lack of appropriate detail and steady growth in numbers.

Librarianship Approach. Instruction in information retrieval for members of many vocations has developed because of the gulf between readers and literature. Probably the best way to bridge the gap is to gear one's self to the reader's viewpoint and understanding. In the first place, the reader is more likely to be appreciative of this approach, and in the second place, he has many obstacles to study requiring effort and attention.

The librarianship approach is one that comes naturally to librarians, but the reason for its adoption should be that it is suitable for students whose interests require both the extensive and intensive use of libraries, the obvious fields of interest being history and literature. It is important that librarians should be seen to be attempting to bridge the gap from the literature side by provision of reader services and bibliographical guides and a ready and able response to requests for help with information needs.

The librarianship approach, by concentrating on library services, is the opposite of that adopted for the basic structure. The following themes which should be present in the introductory sessions of both are:

- aims and objectives of the instruction;
- limitations of libraries within society or knowledge.

Both of these subjects are very important and neither should be taken for granted. The limitations of libraries are no less important than their benefits. If one is able to stimulate use of libraries, there should be reasonable probability of satisfying users; otherwise the result may be disenchantment.

More Extensive Approach. The basic outline of teaching material is a compromise between a number of variables so that it can be used in a wide range of situations. It would be ideal if the instruction in information retrieval could be extended over a greater period of time, eg. back to school days, so that the skills are properly developed. Unfortunately this is impracticable in many situations, hence the compromise. More extensive instruction in information retrieval is to be favoured when it forms part of educational programmes which give relatively small core knowledge but wide-ranging understanding and skills - in other words, which can develop people who are both creative and able to use existing knowledge.

---

[1]ANDERLA, G. Information in 1985: a forecasting study of information needs and resources. Paris, OECD. 1973. 131p.

"An interesting product of such education is Barnes-Wallis, an innovative British
aircraft engineer and scientist and known for a number of developments including
'swing-wing' aircraft. His biographer (J.E.Morpurgo. Barnes-Wallis: a biography.
Longman, London, 1972. pp.63-64.) notes that during Barnes-Wallis' service in
the army: 'There followed for Wallis an episode which varied and turned almost
inside out, the Army's oldest and still current joke. He was summoned by the
Adjutant. "You're an engineer, corporal", "Yes, Sir", "Then go off and design a
sanitary system." Wallis spent three days in the Reading Room of the British
Museum and then designed and built with their aid a vast system suitable for a
thousand men. Of all his engineering feats it remained among those of which he
was most proud."

In this ideal approach the teaching of information retrieval would be comparable with
teaching numeracy and effective writing.

Only a general outline can be given here, but one might expect instruction in information retrieval to start at the level where most children can read a wide range of
material including a children's encyclopedia. At this level some school teachers already
provide a 'finding-out' corner which consists of a few carefully chosen informative
books and basic measuring apparatus. The content of such a collection needs to be varied
to meet particular teaching programmes. Thus if the theme of 'growing food' is integrated
into all facets of the syllabus then there needs to be an adequate collection of books on
the subject, including items specifically developed by the teacher or other pupils.

It is wise at this stage that the class-room collection should not exist in isolation;
the school should maintain a supporting collection, with minimal duplication. In turn the
school collection can be supported by the local library or educational authority, including qualified staff. The actual structure of collections must depend on the particular
situation, but the emphasis should be on small collections of high-value material, possibly with recreational material as well.

Using the material ought to become a natural part of studies and pupils who seek
information from the teacher should be expected to have at least considered using the
'finding-out' corner. Similarly when the teacher's assistance is sought and the appropriate information source is either literature or observation, it may be of greater educational value to guide the student than simply provide the actual information needed. The
principles learned can be: use of literature as a source of information; the interdepen-
dence and use of different collections of literature of different levels and sizes; variations in quality of literature; and use of individual items and index terms.

The next important stage in instruction will precede significant project work or, if
projects are not used, the study of topics in depth. At this point, more use can be expected of a wider range of material, including the libraries available to the community
in general. Thus there will be a change from using known sources of information, such as
a children's encyclopedia, to unfamiliar material which is located using catalogues and
indexes. In short, guides to the literature are introduced. Children can be taught to use,
or even produce, a school-library catalogue. Purchased guides, such as a national bibliography, can be used. It is important that librarians and teachers should cooperate to
ensure that adequate material is available and that necessary skills are 'learnt' prior
to projects.

Later still children may be expected to work on practical studies of various subjects,
including laboratory work and field studies. They should be expected to compare their
results with those produced elsewhere and to give explanations for possible differences.
Ideally, published abstracting and indexing tools should be used.

All this is admittedly idealistic, but such thoughts can act as a focus for developing
objectives. Constraints do exist in both developing and developed countries, and one needs
to think over a broad area but operate within the existing constraints, of which financial
and geographical limitations are important. However, a society creates its literature and
there is no reason why children should not create it too. Such activities can be used by
the teacher to teach and motivate pupils. Inclusion of the results in the library can be
compared with the adult's publishing activities. Similarly classification, and to some extent cataloguing, can contribute to the learning process, while abstracting and indexing
are simply variations of précis-writing.

If instruction in information retrieval is extended to schools, teachers in universities and colleges need to adjust their instruction accordingly. The abilities of students
need to be examined before instruction, which can then start from the accepted level. The
examination can be a simple test followed by remedial instruction, possibly in the form
of programmed texts. The same techniques can be used for postgraduates. The basic course
structure can be retained, but the actual content needs to be adjusted. This first lecture
can be seen as a consolidation of existing knowledge. The value of information can be

taught by a simple review of case studies and basic theory. The identification and occurrence of information needs will be a more important subject than the importance of information retrieval. Instruction in the principal sources of information can emphasize variations in particular needs and limitations of the various sources. The levels of bibliographical tools can be covered very briefly, as the students are probably aware of these already.

<u>More Intensive Instruction</u>. One would like to see the development of information-retrieval skills over an extended period of time. Countries with developing educational systems are encouraged to consider the importance of such skills, but are often forced to develop them intensively in a fairly short time. This problem may arise from a short allocation of teaching time or from the short duration of the complete course. Handouts and additional reading can be useful for short intensive courses; also emphasis on the practical aspects. The measure of the achievement of the course is the extent to which students subsequently use information-retrieval skills. Even a very basic skill can be fruitful, provided that the student recognizes its limitations. However, the teacher must remember that students can only accept a limited amount of new knowledge in a certain time. We suspect that many people over-estimate this amount.

Finally, it cannot be stressed too greatly that intensive instruction is of limited value, especially where the available library and information services are operating in restricted circumstances. In a developed country the information services may make up for deficiencies in instruction. Also, information demands on the practising scientist or engineer may be less because he has colleagues to consult and his current work has similarities with earlier experiences. In developing countries, the engineer or scientist may meet difficulties which are novel to him and his sources of information and assistance may be more limited and remote. It is therefore important that he can use the available information resources well.

5.2.5 <u>Differences Arising from Locale</u>

The basic course outline in this Guide can be used in both developed and developing countries. Differences between them call for a brief study of local situations. For example, developing countries are probably bringing one or more features of their societies to levels already achieved by more industrialized countries - but at some cost of resources and expensive errors or trials. By using information which is often available in the published literature, they may be able to attain the same level of achievement at less cost. It is worth considering such a possibility as being within the topic 'the value of information'.

A second difference, which can be substantial, is the accessibility and quantity of literature and recorded information within a country. It can affect the relative use of the main sources of information, particularly oral and recorded sources.

It is useful to modify the basic structure according to particular circumstances. Possible variations, covered in this Chapter, are:

- subject background,
- educational level,
- theoretical or practical bias,
- librarianship or information-science bias,
- extensive or intensive instruction,
- newer emphases.

Subject and educational differences are usually inherent in the local situation and need to be accepted by the teacher. In contrast, a particular bias, be it a theoretical, vocational or librarianship, is a matter of judgement for the teacher. Finally, the virtues of an extensive or intensive instruction and the inclusion of newer emphases based on the developing situation are a matter of conjecture. The teacher needs to view these factors in terms of his main objectives and within existing limitations.

5.3 OTHER IMPORTANT TOPICS

There are a number of important topics which can be included in the basic structure where appropriate qualifications exist. The qualifications are: accepted level of motivation and knowledge of students; ample amount of time available. The existing knowledge of information retrieval is growing and certain assumptions about it may need to be changed.

5.3.1 <u>Occurrence and Identification of Information Needs</u>

This topic can be included at postgraduate and other advanced levels, with highly motivated and practical students.

The occurrence of information needs which can be related to a new situation or event has been referred to earlier as a 'nova'. (See Section 5.2.4 - Theoretical Approach).

For engineers, at least, 'novae' result in information needs as follows:

- New project, job or task: awareness of current understanding of the pertinent
  subject area.
- Awareness of new processes, products, ideas, techniques, etc. which influence
  current work.
- Working information for the execution of assigned tasks, eg. data and product
  information.

These three needs relate to the three distinct types of search - retrospective, 'current-awareness' and quick reference respectively. In practice mis-matches between information needs and resources cause inefficiency of information flow. For example, text books and conference proceedings may be treated as the same by a library but they are used to satisfy different needs. One suspects that conference proceedings, though not neglected, may not be fully used in relation to their value.

If the occurrence or frequency of information needs is plotted against time for an individual engineer, the resulting graph is likely to be an irregular series of peaks of differing heights. The peaks are associated with 'novae'. Thus the engineer can sort out his needs whenever his circumstances change and, provided he takes reasonable action, can be informed of new events elsewhere related to his work.

Instruction can be based on these lines, with the possible addition of particular cases for practical illustration, these cases being selected according to background of the group of students. A simple case with clearly defined elements is the Saturn $\overline{V}B$ which has already been referred to earlier in this chapter.

The fuel tanks for Saturn $\overline{V}B$ rockets were pressure-tested using a different chemical to that intended as the fuel. Although the two chemicals were believed to be similar, the use of another chemical was a departure from the planned application and constituted a 'nova'. 'Novae' such as this are not unusual and the consequences are frequently negligible. Engineers ought to be alert to information needs arising from 'nova' but, because serious consequences are infrequent, the needs do not always receive proper attention. The sensible approach is to consider the seriousness of possible mischances and adjust time spent on information, as appropriate. It would be foolish to do so on the apparent probability, or improbability, of mischances, since they are unpredictable as the following simplified account of an actual event illustrates well.

"The jib of a mobile crane collapsed during a lifting operation and as a result
several people in a passing bus were killed or seriously injured. The subsequent
enquiry revealed an almost unbelievable series of events with associated 'novae'
and information needs. It is possible that attention to any one of these events
could have prevented the collapse. The crane had been modified (nova 1) to keep
the overall length of the vehicle within that permitted by new traffic regulations. The modification was a hinge piece in the jib. Although it may have been
suitable in idealized circumstances, a loose fit or undue play could result in
increased stress at the root of the jib. The designers did not check whether
their ideal circumstances existed. Secondly, the resultant hinge included parts
remade (nova 2) using an incomplete set of plans, with the result that they
were thinner and weaker than intended. Thirdly, the crane jib was overloaded
when making the attempted lift, and this was aggravated by a slope (nova 3) when
the limitation of operating on a slope was not appreciated. Fourthly, the crane
was operating with the overload indicator and cut-out switched off (nova 4).
Finally, vehicles were permitted to pass on the nearby road when the hazardous
manoeuvre (nova 5) was attempted."

It is important to stress that new events or departures from planned and accepted procedures are likely to produce information (and associated communication) needs. As the above example shows, neglecting these needs in engineering fields can be very serious. The design, testing, production and maintenance fields are particularly important.

This leads to identification of information needs. If information needs are prominent when 'novae' and departures from accepted practices occur, one needs to be alert at these times and systematically examine what one is attempting to achieve and what methods are being used. The focus of attention can be on the novel elements. In more routine matters one often accepts value of checking calculations and duplicating or triplicating experiments. Similarly a checking of information used ought to be a matter of routine. This checking could simply be - "Where did the value come from and is this acceptable?" All this assumes individual action but, where projects involve teamwork, there ought, in an ideal world, to be contact and use of an information service. In significant projects an information officer ought to be considered as a member, even if part-time, of the team. This would seem to be expensive but some cost calculations have been attempted which show that it can reduce total costs.

5.3.2 Control Factors in Information Needs

The wide variety of factors in information retrieval makes for substantial differences from search to search. Although these differences make searching interesting, they also create difficulties in instruction.

Control factors are an important part of information retrieval in that once an engineer or scientist has acquired information retrieval skills they need to be used wisely. For example, an individual could spend an excessive amount of time seeking information once he is alert to a 'nova'.

More work is needed to establish the control factors precisely, but the following warrant inclusion:

- estimated cost/benefit of satisfying individual information needs;
- probability factors;
- quality factors;
- available resources;
- receiving and digesting information.

The estimated cost/benefit of satisfying individual information needs is simply a recognition of the variation in potential value of specific items. Thus the user should be aware of the significance of the required information and what amount of time, effort, and money is justified in collecting it. Experience and better understanding should help to reduce costs and increase benefits. Cost/benefit analysis of total information provision is a matter for information scientists; they have had difficulties in quantifying the value of information since it is often obscure, whereas costs are very clear and often seen as substantial. Some studies indicate that there can be very high returns on the investment in literature and information, but that such cases occur relatively infrequently. Apart from these, there is a possibly small but continuous return on investment of time and money in libraries and their use.

Because the value of specific information is difficult to determine, one has to assess the second factor - probability - which arises because one does not normally know if the information is actually available. A useful analogy is seeking submarines in a large ocean. One searches without knowing whether the objective is actually there, but techniques can be developed so that one looks in the most probable area first.

The quality factor affects information searching in a number of ways. One easily appreciates variations in the quality of primary literature (books, periodicals, etc.); but there are also variations in bibliographical tools and other forms of secondary and tertiary literature. This is an advantage in countries and organizations with limited financial resources because about 90 per cent of information needs can be satisfied by 10 per cent of the available literature; a small collection of material can produce a high percentage of satisfied information needs. It is not suggested that all library collections should be small because some highly beneficial items of information may come from more obscure items of literature. Thus a single large national library with substantial holdings can co-operate with other librarians and so permit local collections to be smaller than if they had to be relatively self-sufficient. On a more individual level a scientist or engineer (including an engineer working in the field) can usefully have one or two text-books, periodical articles, etc. at his immediate disposal, but rely on other sources for the rest.

Another facet of quality is the 'confidence limit' of pieces of information, particularly statistics and data. For example, the accepted value for the speed of light has changed over the past fifty to eighty years. In practice, if two authoritative sources of information give similar values or data, they increase substantially the confidence in, and acceptance of, the values. Confidence is further increased where three or more values are similar, but the rate of increase is less with each additional source. Organizations which collate available data from all sources serve a very useful function, but more work of this kind needs to be done.

Available resources are a factor discussed in far greater detail elsewhere in the Guide. They include librarians and other intermediaries.

The final factor is reception and digestion of information. Many practical problems reflect not so much a lack of information but an excess of information which often swamps or camouflages the information needed. Different people have different digestion rates for information and need to take this factor into account when considering their information needs.

5.4 EDUCATIONAL OBJECTIVES AND CONCLUSION

The basic subject content of the introduction to information retrieval ranges from very fundamental facts to almost philosophical concepts of information transfer. This is necessary in order to serve the wide variety of users who need instruction in information retrieval. The educational aims remain common to these different categories and can be simply stated as objectives after receiving initial instruction. Students ought to be able to describe:

- the role and value of information retrieval (or library use or information acquisition) within their vocation;
- the content of the library or information centre as a source of information compared with other principal sources of information;
- the limitations and obstacles in using libraries and information centres and some of the ideas and concepts used to overcome the obstacles.

In addition the students should have been motivated sufficiently for them to intend using the library to satisfy some information needs. There are also educational objectives for the teacher. The course may be the first time that he has met the students and it is important that he should become aware of their background and attitudes, for example by discussions with appropriate people. It is of some use to have the initial session timetabled after a free period, since discussion with early arrivals can provide useful impressions. This gaining of familiarity helps marginally and it is the sum of marginal improvements which make for professional teaching.

5.5. CHECKLIST USING THE APPROACH GIVEN IN THE GUIDE

1. Have you established the approach which you will use? The main choice is between an information (resource and need) approach and a literature approach.
2. Does your introduction put the instruction into content of studies and subject field? Have you clarified the circumstances in which there is a need to use information retrieval/library skills.
3. Do you need sources of information outside libraries and literature?
4. Have you indicated the limitations (and the scope) of information-retrieval skills?
5. Will your students/audience be able to resolve difficulties and choose relevant options which may occur?
6. Has your presentation/lecture/seminar taken sufficient account of the viewpoints of your students/audience with respect to subject area, educational level, and profession?
7. What changes in performance do you expect from your students/audience?

6. Course content 2—Basic methodology

Having established the fundamental principles of information retrieval, attention can be given to the basic methodology. This methodology, together with closely associated practical exercises, is the more important part of the instruction. It is also a difficult part in that timing, group size and differing attitudes all influence a particular situation, and the details of instruction are likely to change accordingly.

Almost any instruction in information-retrieval techniques must include consideration of search terms, search tools and search strategy. These three facets can be shown in a variety of ways and in different orders, and they may be represented as follows:

Figure 6.0: The Facets of Basic Information Retrieval Methodology

For the initial introduction to users the use of the word 'search' is useful, although later the three facets will appear under different headings: search terms being related to 'indexing and classification', search tools becoming 'bibliographic tools' and search strategy becoming 'steps in literature searching' or even 'case studies'.

The facets of basic searching obviously interact. Thus different search terms may be needed for different bibliographical or search tools. Users need to be aware of this interaction. It is also desirable to minimize the amount of factual information to be learned by users. The objective is to choose the right strategy, tools and terms, with the minimum of effort and frustration. A number of working methods are used in pursuing this objective and it is suggested that consideration is given to a combining two or more of them.

6.1 WORKING METHODS

<u>Counselling</u>. The suggested practice is to develop both formal and informal relationships with users, so that formal instruction is followed by users visiting information officers to discuss their information needs. On such occasions the users are normally motivated and receptive, and both general and specific skills can be developed.

The balance of 'formal' lectures and 'informal' counselling will depend largely on factors such as the range and quality of local resources. It will develop naturally to some extent but nevertheless should be taken into account when planning. For example, two comparable university libraries may plan on giving their students six hours instruction in information retrieval. The first may plan two hours formal instruction for each of three years (year one - arrangement and use of the library; year two - fundamentals and basic methodology of information retrieval; year three - searching for information for final-year projects and other matters such as current awareness). In addition, 10 per cent of the students may seek advice; so receiving further informal instruction for about one hour. The second library may plan for three hours formal instruction in the first year with a further one to three hours informal instruction initiated by project work and agreed with course tutors. This latter approach is, of course, the one to be followed with both postgraduate students and practising engineers and scientists.

Practical Exercises. The teacher can initially set exercises based on what he believes to be the basic information needs which the student ought to be able to satisfy on his own. This is an important part of the instruction since it brings together at least two of the facets of methodology (search tools and search terms) and in particular seems to be the main method of teaching the practical skill of choosing search terms.

However, local circumstances usually impose limitations, such as the depth and arrangement of stock and the number of users, and the result is often a compromise. Some exercises may be completed by luck (or serendipity) rather than by the application of intellect; and a lack of reality often results from attempts to develop representative and fair questions needing a particular level of skill.

Case Studies. The advantages of case studies are that general principles can be clearly illustrated and thus become more meaningful. Indeed group participation can be used to advantage. Case studies can take the form of a tape/slide presentation, but choice of subject can be difficult and good presentations take considerable time and effort to produce.

The limitation of case studies is that a single example cannot be representative of the whole range of information needs. Different subject groups may require different cases, and these needs can be more demanding, especially if films or tape/slide presentations are made. The subject of the case can distract from the principles of information retrieval. Finally, some groups may not respond to participation, possibly because they have a high number of time-tabled lectures or because they are not accustomed to variety of teaching methods.

Printed Guide or Notes. There is a lot of purely factual information which can be provided for users in the form of a printed guide or detailed notes. However, it is essential that these are used to supplement the instruction given in the lectures and practical exercises (like the example in Appendix 2).

Flowcharts and Search Programmes. A piece of information is sought by carrying out a series of relevant steps. It is possible to create a flowchart which initiates students to a wide range of searching procedures. The steps in a flowchart do not have to be learned, but can be read like a map. The flowchart is likely to be fairly complicated and a number of simpler charts may be necessary. The result can be assembled as a programmed guide. It is possible to produce a package for a teaching machine, particularly if a computer or simple display plus 'memory' is used.

The advantage of flowcharts and search programmes lies in teaching search strategy and how it correlates with the other facets. Ideally, the material ought to reflect the different approaches that can be used and ought to be available in the library as part of its broad guidance. The disadvantage of flowcharts is that they substitute written instructions for an intellectual process, this having been carried out by library staff. The result is that users in another library, or who do not have the flowcharts, may be unable to use a satisfactory search strategy.

Flowcharts and search programmes should be regarded as an optional working method for guiding users on search strategy, particularly on the decision processes. Given ample time, students can learn to draw up their own flowcharts and thus gain some understanding of the factors involved.

Emphasis on Decision Processes and Problems. If flowcharts are difficult to prepare in detail, an alternative is simply to emphasize the decision points and supply or indicate the alternative bibliographical tools and sources. In other words the flowchart is reduced to key points. This is useful because the students may not appreciate the existence and nature of different routes to acquiring information.

6.2 TOPICS COVERED (LOGICAL APPROACH)

For students receiving instruction in lectures plus practical work, the lectures might be as follows:

> Basic technique - choosing search tools
>   - establishing a search procedure
>   - choosing search terms
> Search variables
> Case study
> Briefing for practical work.

One illustration showing the three facets has been given in Figure 6.0. Alternative arrangements are possible, as shown in Figure 6.2. A stylized representation of the facets can be useful and used repeatedly.

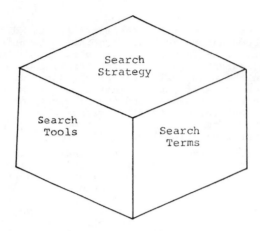

Figure 6.2: The Facets of Basic Information Retrieval Methodology: Alternative Arrangement

An example of a detailed handout to be used in conjunction with this and/or later lectures is given in Appendix 2.

6.2.1 Introduction

A brief statement as follows should suffice:

"Today's lecture is a crucial one. We are concerned with the actual methods of finding information in a library.
There are three facets - search terms, search tools and search strategy - and these facets are repeated in your handouts and on the blackboard.
We will take each one in turn. It does not matter which is first - but it is preferable to start with search tools and finish with search terms.
Having considered each facet in turn we will consider some variables which affect each search and how the facets come together by looking at a case study.
As usual we will conclude with a briefing for the practical work."

6.2.2 Choosing Search Tools (i.e. finding or bibliographical tools)

There are several different types of tools, such as book-lists and periodical-article lists (abstracting and indexing journals).
Terminology is a problem, abstracting and indexing journals being a classifical example. Another problem is that different libraries do not put the different types of search tool into the same categories. Indeed some search tools (which are in fact lists) cannot be categorized at all. The difference between a bibliography and an abstracting journal may be vague and may depend on the method of purchase - book order or subscription. As a result, the teacher must provide a simplified description which permits a degree of self-sufficiency, ie. the user may use the guidelines for 70 per cent of his needs. The remaining needs can be regarded as 'problem needs' and separate guidelines can be given, working from first principles and after consulting librarians and information specialists.
The lecture may continue with a consideration of four main groups of search tools as follows:

> **B**ook-lists
> **A**bstracting and indexing journals
> **R**eference books
> **B**ibliographies and guides to the literature.

The word BARB can be used as a mnemonic.
It is now necessary to discuss the nature of each category of search tool:

Group 1 - Book-lists. The name is self-explanatory. Most users will be adequately served by two or three different lists from a fairly small range of choice The main problem with book-lists is that they do not give any detailed indication of the contents. The information which assists choice is minimal.

For most scientists and engineers the following lists should be adequate; they can be included in a handout and used for practical exercises:

1. library catalogue;
2. published book-lists for current topics, e.g. a 'books-in-print' publication or a national bibliography;
3. published book-lists for supporting studies, e.g. a published library catalogue.

Types of needs determine the precise choice and range of coverage; they will have been considered in the educational objectives. The average student should be able to find and use the chosen two or three book-lists for his subsequent work and to locate them by name or description in another library.

The library catalogue is based on the library's own collection of books. Its use should be straightforward and obvious. Normally, it is in two parts: a name (or author) catalogue and a classified (or subject) catalogue. For a classified catalogue the means of translating subject terms into classification codes will be needed.

A domestic list of books-in-print, if available, supplements the library catalogue. Books-in-print are publications which list books currently available from the book trade in a particular country or language. Most differences between a library catalogue and a books-in-print publication are fairly obvious and relate to priorities - the urgency of the need, the adequacy of the collection for the topic, the quality of the catalogue, the currency of the topic. Other differences are less obvious.

Many countries produce a national bibliography, which is a catalogue listing of books published in that country. Most librarians and many information officers will choose a national bibliography in preference to a books-in-print publication. This is because national bibliographies are firmly established as useful publications for librarians, especially cataloguers, but books-in-print publications are produced essentially for booksellers and others in the book trade. In spite of the usefulness of national bibliographies, books-in-print usually have a significant advantage for users; they are quicker and easier to use.

Other countries' books-in-print publications and national bibliographies may be useful in particular cases, eg. in a country with limited publishing programmes or using the same language as another. Again, some studies are relevant to another country. One may well be interested in sewage treatment in France, in which case the French books-in-print or the French national bibliography may prove useful. The key problem is the name and location of the nearest copy - a problem on which specialist advice can be sought from librarians.

Some libraries, notably national and other large institutions and those covering certain specialized fields, have catalogues which are published and available for purchase. If these are available, they can often provide a quick and relatively easy method of noting what is published. For social sciences, humanities and arts some of these published catalogues are more useful than books-in-print publications and national bibliographies, which are often based only on current or single-year publications.

Users may not need to use publishers' catalogues often, except in special cases when libraries can use them to provide the easiest and fastest method of locating details of a book.

Group 2 - Abstracting and Indexing Journals. These are essentially guides to periodical articles, but certain services specialize in other forms of literature such as news items.

Some may be produced locally for specialized purposes, but the majority are commercial publications covering a specific subject area. They may be fairly broad in scope, like Chemical Abstracts, or narrow, like Zinc Abstracts. Guides to abstracting services are also available in many libraries, for example that published by the International Federation for Documentation.[1]

Abstracting and indexing publications are key tools in information retrieval for science and technology because they provide details of periodical articles and other literature. A high proportion of the time allocated for searches should be spent using this material.

There are several thousands of these publications throughout the world and individual libraries may subscribe to several hundred. Choosing the appropriate abstracting journal can be a problem. However, many of the abstracting journals are very specialized and in many cases a group of students can be provided with a list of the major abstracting services. (Appendix 2 contains a list which can be useful to English-speaking students, but a similar one can be produced to embody abstracting journals in other languages. Some countries have centralized the production of abstracting journals under a general title with numerous sections for different subject fields. Typical examples are Bulletin Signalétique in France, and Referativnyi Zhurnal in the USSR; a note may be necessary to indicate this type of abstracting journal.

[1] Abstracting Services (2 volumes), The Hague, FID, 1969.

For groups with a common subject interest more specialized abstracting journals can be added to the list, as appropriate.

The difference between an abstracting journal and an indexing journal is that indexing journals, like book-lists, do not inform except to list material which can be consulted; whereas abstracting journals have notes giving some details about (or from) the listed item. The quality of abstracts can vary considerably, however.

Group 3 - Reference Books. Any book may be used as a reference book, for the reader may simply wish to verify, obtain, or better understand a fact or small piece of information. Some books are particularly designed to bring convenient groups of facts together in a single work. Many libraries keep such books together in a reference collection, which serves to some extent as a distillation of data and factual information gathered from many sources. Consequently the reference collection is often the place to acquire day-to-day facts and data or to start a new, substantial search. When this is so, it is desirable to be able to use the collection quickly and easily. Arguably, more library processing time could be spent on such material. As it is, library reference collections may contain from several hundred up to several thousand works, and these numbers can make proper and effective use of the collection difficult. Experience with practical exercises suggests that students have difficulty in identifying and locating appropriate reference works.

The technique for choosing (locating) reference books may vary according to circumstances. The important needs are to check the local situation and to encourage or show that more than one approach can be used. It is suggested that consideration should be given to the following approaches:

1. A library-produced subject index to the reference collection. These exist in only a very few libraries, and it is unfortunate that they are not more common. Although such an index can add to the cost of library processing, the reference collection is considered relatively important and is heavily used for both brief and detailed searches. Mechanized processing systems may permit easy compilation of such indexes, and manual systems, possibly based on titles, can be produced quite easily.

2. Browsing appropriate sections of the collection on the shelves. This is advantageous when the contents of the collection are restricted to use in the library only. One of the advantages of browsing is that the size and content of works can be seen, and partly remembered reference works may be easily identified, since many people can recall details such as the colour of a volume! Disadvantages include the limitations of the classification itself, and the fact that pamphlets tend to be overlooked. It should be remembered that classification schemes often make provision for reference books by form, so that encyclopedias, dictionaries, language dictionaries, timetables and maps may each have a separate classification number. In practice, there are likely to be at least two alternative classification sections - subject and form. Three is common - subject, form and geographical - and this does not take into account the generic/specific scale, eg. electronics information may be found under its own class, electrical engineering, physics, or even technology or science. It seems advisable, therefore, to browse through several sections: some theoretical studies of browsing, based on the analogy of seeking submarines, indicate that browsing several small scattered sections is likely to be more profitable than searching a single large sequence for the same amount of time.

3. The third and most common means of location is a reference librarian or similar specialist. This individual's knowledge of reference material, built up over a number of years, is extremely useful; it is usually buttressed by published directories of reference books.

4. The user himself can use the published directories of reference books. These are fairly extensive but suffer from soon becoming out-of-date, and are not as specific as most users would desire.[1]

---

[1] The notable works are:
WALFORD, A.J. (ed.): Guide to reference material. 3rd edition (3 volumes in progress) London, Library Association. 1973
WINCHELL, C.M. : Guide to reference books. 8th edition (with later supplements) Chicago, American Library Association. 1967

5. The subject guides to literature also list reference books appropriate to the particular field. One should note that reference collections are developing and changing, whereas many subject guides are dated and vary in quality.

Group 4 - Bibliographies and Guides to the Literature. Although both bibliographies and guides to the literature are often incorporated into one of the other collections of search tools, it is helpful to discuss them separately with users.

Originally, a bibliography was a list of books, but it is now extended to cover any list of references, usually with a common element - a particular topic or some other feature. If a bibliography exists on a topic which is being sought, then time and trouble can be saved. For the information user an annotated or extended bibliography is particularly useful. Ideally, a literature search ought to incorporate a few minutes' checking whether a bibliography is available. Cross-references and subject headings sometimes indicate its existence and location. For scientists and technologists this is the ideal situation because bibliographies on many search topics are infrequent. An experienced searcher can sometimes anticipate the likelihood of bibliographies existing. The topics are wide and include topics of common or current interest such as pollution, holography, fatigue of metals, lasers, noise and machine tools.

Bibliographies appear to play a much greater role in social sciences, humanities and the arts. This is not surprising in view of the fact that literature in these fields remains of interest and relevance for much longer periods of time. Linked to this, many topics extend to before the period of substantial growth of present-day finding tools. In practice, finding tools that list bibliographies are available.

The teacher needs to examine the local practice with bibliographies and decide what guidance needs to be given to information users.

Guides to the literature are different. It is obviously important that they are easily located in a library, particularly one lacking efficient information and reader services. Unfortunately, in many cases they are not easy to locate. Those judged to be out-of-date, superseded or of poor quality may be best discarded.

The teacher should examine local practice and judge what guidance is needed by information users. This is not always easy because guides to the literature are a significant topic themselves, but the range needs extending to cover notable groups and deficiencies.

Basically, there are two types of guide to the literature. One is concerned with literature of a particular subject field and the other with particular types of finding tool.

Subject-based guides to the literature are useful to information users in fringe subjects who are uncertain of the appropriate finding tools; and also for providing details of the more unusual finding tools for specialized information needs and categories of literature.

Guides to particular types of finding tools are useful if kept current, and libraries and information centres can usefully consider compiling lists for consultation by their own users when published guides are found wanting. The important thing is that the information user should know which finding tool to use with the minimum of time and mental effort.

6.2.3 Establishing a Search Procedure

Although it is impossible to provide a simple series of instructions to cover all searches, there are common types of search which can be used for the design and basic procedures. Information needed to enhance an individual's knowledge of a topic may be described as a retrospective search, reference retrieval, state-of-the-art-search, statement search, or literature search (each having slightly different meanings). Attempts to be precise about distinctions and discuss 'retrospective reference retrieval' can confuse students and should be avoided.

There are a number of approaches to describing a search procedure. One method is to state a number of steps as in the following example:

Six Steps in a Literature Search

1. Define the information need.
2. Seek clues or leads.
3. Decide search policy (choose search tools and search terms).
4. Note all bibliographical details of potentially useful references (author, title, etc.).
5. Look particularly for bibliographies or review articles that may be used as a basic reference source and then up-dated with more recent material.
6. Having found the references, obtain the material and assess its relevance.

This list of steps is attractive and incorporates significant aspects, some which might be overlooked by students. The main weakness is the likelihood that it cannot be recalled accurately from memory. (The reader can test this now by closing the Guide and attempting to write the steps down on paper.) If the steps can be consulted by reference to a handout, this problem is avoided. However, if there is a need to document the steps in detail, they can be elaborated or written in the form of a flowchart as in Figure 6.2.3. Better still is the list of steps that are both simple and elaborate; simple in that the steps are sensible and obvious and therefore can be recalled; elaborate in that useful detail can be obtained from a handout. This dual approach is attempted in the handout in Appendix 2.

Three steps are given in the developed strategy, and they are based on the problem-solving approach: define objective; plan; carry out plan (adapting as necessary); and, possibly, review and supplement the results. Mnemonics are possible (OPENS - objective, plan, execute, note progress and supplement) but are probably best left to the enthusiast: BAR OPENS might have certain advocates. But an individual user is more likely to develop, remember and use a search strategy which is appropriate to his own needs. In addition, he will adapt the strategy more easily to local circumstances.

Another approach, which can be more satisfactory for small and intelligent groups, is for the teacher to produce a listed search procedure during the lessons. This takes time and teaching skill but the result reflects the thought processes of the group and users may subsequently repeat it to some extent in real situations without recourse to lecture notes. The whole matter is crucial to learning information-retrieval skills and deserves the necessary time and attention.

Establishing a search procedure in the classroom can be done in a number of ways. A case-study can be used, but it is then particularly useful to check that the procedure is suitable for other cases. A question-and-answer session can be used, but there is a risk of excessive teacher influence, which would spoil one of the major advantages of the approach. When supervised practical sessions are to be used later, briefing can be combined with establishing a search procedure. The approach still has much potential for further development and teachers are advised to consider this.

6.2.4 Choosing Search Terms

Of the three aspects of methodology, the choosing of search terms is likely to be the most persistent cause of difficulty.

The best way of learning about the principles and problems is by exercises, particularly under supervision. In the lecture room, difficulties and examples can be discussed, for example the example of seeking information on the Stirling Heat Engine.

Search Tool	Search Terms Needed
Applied Science and Technology Index	Heat engine
British Technology Index	Stirling heat engine
Engineering Index	Air engine

In this example, cross-references would greatly help the searcher, but they are not always available even in good-quality abstracting and indexing journals. Similarly a searcher wanting information on the bulk density of certain minerals would have to find the index term 'weight-volume relationships' in one reference book and 'aggregates' in another.

Examples such as this cause surprise to most library users when they first encounter them. Most people are completely unfamiliar with the nature of indexing. Thus, instruction in handling indexing problems needs attention.

"I have shown you these examples of the problem of choosing search terms because, unless you have experienced these difficulties, you don't know they exist. In many cases, the alternatives appear to be unbelievable. What can be done about it?

First of all, the cause of the problem. It is the 'concept to words' problem discussed in the fundamentals session. Remember example of the perfect aeroplane. In practice, the indexer is a different person in a different place at a different time with a different view.

Secondly, action being taken to improve the situation. The result is the co-existence of different types of indexes, because we have not one difficulty but many difficulties, associated with linguistics, costs and amount of available time.

Thirdly, and most important from your view, what action can be taken by the searcher? Basically, he has to do something along the following lines:

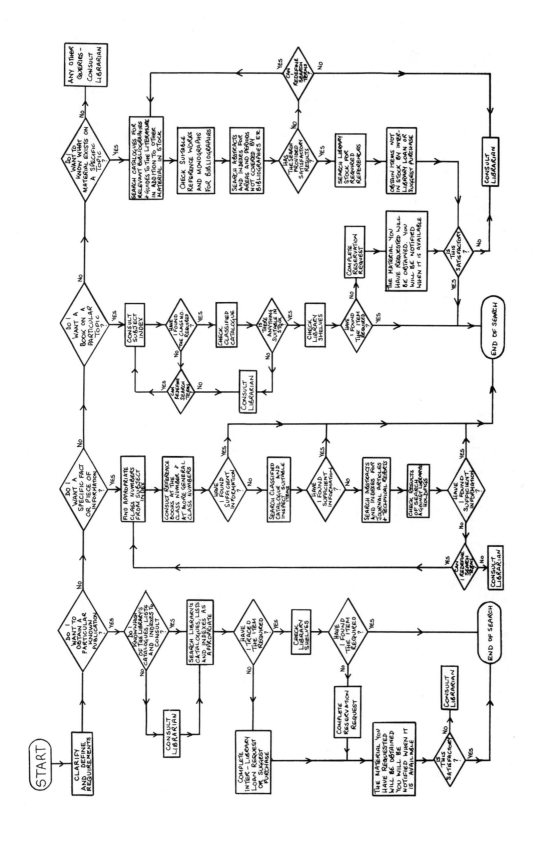

Figure 6.2.3 : Flowchart of Search Procedure

1) Define in his own words what he wants;
2) Consider any words other people might use;
3) Note any synonyms;
4) Systematically check each possible heading in each search tool until relatively satisfied;
5) If not satisfied, try more general terms and be ready to search under both specific and general headings;
6) Always review the results and look for reasons for unexpected or unsatisfactory results.

Each search for information in the literature is different and some of the main variables are listed in Figure 6.2.4. It is these variables and the basic decision associated with:

> search terms,
> search tools,
> search strategy,

that make the process an intellectual rather than a routine matter.

The material expressed on Page 71 supplements and modifies the basic content according to the local circumstances. Reflection is worthwhile at this stage, and one obvious possibility is to bring together the three facets in a typical situation. A case study is one useful method."

6.2.5 Case Study

When describing basic techniques of information retrieval it is useful to illustrate with realistic and typical cases. In addition, it is useful to combine in a more extended case study the various aspects of search techniques. The case study can involve role-playing or another form of participation, or a prepared presentation.

The case study and its method of presentation will need to vary according to circumstances. Firstly, consideration can be given to a prepared audio-visual presentation such as a tape/slide set or film. Such presentations are attractive, ease pressure on the teacher, and can be used over and over again. In addition, this part of the course, which is concerned with communicating skills in information retrieval, contrasts with the part that develops group interest and deals with basic principles. It can often be easier to maintain the stimulus with audio-visual presentations. However, producing them requires much time and effort if they are to yield satisfactory results, so much so that academic libraries in the United Kingdom have co-operated in producing a range of them. Each presentation may have been developed to meet a specific need and it will be found that certain subject groups are not sympathetic to the topic used. If the presentation has been developed outside the course, the intended use may be different from the requirement of the individual teacher. Add to all this the need for equipment and darkened rooms plus the risk of equipment failure, and even enthusiasts may be deterred from using the technique.

However, experience shows that they are useful when suitable items are available or can be prepared. It is suggested that they give superior results to overhead projectors and commentary, and are more flexible and more easily adapted and shown than cine-films. The picture quality is superior to television recordings. The equipment in a lecture room should be set up and tested prior to the lecture or be the responsibility of a separate operator. As to the presentation itself, a duration of fifteen minutes is ideal for the case study, with twenty minutes as the maximum. Presentations which do not fulfil the teacher's needs should not be shown in full in the lecture. If they are thought to be useful but conflict with the teacher's needs, they should be made available for separate viewing, including viewing by individuals.

The teacher should prepare alternative material or approaches in case a suitable room or equipment is lacking or the equipment fails.

For large groups role-playing and personal involvement can be difficult and unsatisfactory. A set of charts, prepared scripts for selected individuals, or overhead projector transparencies may be held in reserve when planned presentations fail.

For small groups formal prepared presentation may be considered to be more restrictive than an informal active case study. Thus, serious consideration should be given in such cases to interactive involvement of the students. Charts and prepared scripts can help to initiate involvement. The cases used can be more numerous and flexible than a formal presentation, but may take up more time than would normally be justified for a case study. The problems in role-playing are described later.

ENQUIRY PARAMETERS	TREATMENT FORMS	ORGANISATION OF KNOWLEDGE PARAMETERS
Reason for enquiry (purpose to which data will be put) Degree of detail needed Specific fact versus subject area 'Anything' on subject versus 'as much as possible'	Bibliographies of bibliography Bibliographies, publishing lists, catalogues Dictionaries, glossaries Encyclopedias Handbooks, data books Standard works State-of-the-art surveys, literature reviews, 'Advances in ...' series	The various subject contexts of the same topic (due to (1) different applications of ideas or things (2) 'separating' effect of classification schemes) Fringe subjects Analogous ideas or things Classification, and the relation between general and specific groups
GENERAL PHYSICAL FORMS		
Books Periodicals Documents Pamphlets Trade Literature Patents Microforms:(of any of the above)	Symposium, etc. proceedings Textbooks Monographs Theses Directories: town, personal, trade Atlases, maps, guides	**SOURCE PARAMETERS**
		Number of sources involved and 'scatter' of information throughout them Type of abstract or index searched: comprehensiveness of subject covered, and titles covered; publication delay Variations in terminology and arrangement of indexes
PRESENTATION OR POINT OF VIEW	Statistics	
Theory, research Practice: development, production, materials Economic, commercial Use, operation Installation, equipment, maintenance Personnel Organisation & management Social, ethical, public relations points of view Instruction, education: elementary, intermediate or advanced; technical or popular presentation Law History	Tables Abstracts Contents lists Articles News items Adverts., trade catalogues Research reports Specifications, standards Annual, etc. reports Official publications	Authority Whether up-to-date Type of literature source: firms; research bodies, government or private; societies; institutions; universities, colleges; individual authors; commercial publishers (e.g. of trade journals) Whether available locally or to be obtained from external source Type of library searched: academic; public; industrial; research bodies; societies, institutions
	TIME, PLACE & MEDIUM	
	Date or period Country Language	

Figure 6.2.4 : Searching the Literature - Main Variables

6.3 MODIFICATIONS AND CONSIDERATIONS ARISING FROM DIFFERENT CIRCUMSTANCES

6.3.1 Subject Differences

The search techniques and main variables tend to be similar for most subject groups, particularly those in science and technology. Consequently the notes are brief except for notable features and opportunities.

Science in General. Little modification, if any, is necessary. Existing audio-visual aids such as films and tape/slide presentations may be useful.

Mathematics. Little modification, if any, is necessary. English-language abstracting and indexing journals tend to be limited for this subject in comparison with others.

Astronomy. Many of the finding tools are specific to the field. It is probably worth listing some of the more important, if published lists are not available.

Physics. This subject is well served by abstracting journals - both in general and on specific topics.

Chemistry. This subject is very well served by finding tools, some being so notable that they overshadow the remainder. The practical work, if not the case studies, should demonstrate that the alternative sources are still necessary. Because of the comprehensiveness and size of some of the notable tools, such as Beilstein's Handbuch der Organischen Chemie and Chemical Abstracts , separate guides on the use of these particular tools may be helpful.

Geology. Little modification is necessary. The case study ought to involve a wide range of finding tools. Many scientists and engineers tend to be too objective and overlook alternative lines of thought. For example, there is a tendency to use Chemical Abstracts for geochemistry topics and not other aspects of geology. The coverage of Chemical Abstracts is far more extensive than its name implies.

Life Sciences. Little modification is necessary. The subject field is well served by bibliographical tools.

Technology and Engineering. Little modification, if any, is necessary. Existing audio-visual aids such as films and tape/slide presentations may be found to be useful.

Medicine. Case studies ought to deal with the reliability of information. Reliability is otherwise a topic left until data are discussed.

Management. Unfortunately this subject field is not particularly well served by abstracting journals. It is treated partly as a technology and partly as a social science. It is worthwhile, possibly, to mention the particularly useful bibliographical tools and to stress the intermediate nature of the topic.

Social Sciences. The differences between the social and natural sciences become apparent in the practical search techniques. Although most of the principles still apply, the outlines given earlier in the chapter do not appear to be as relevant as some alternative approaches. The problems and the needs are different. For example, ephemeral material is more important, as are official publications and news material. It is difficult to identify a basic technique and the subject is really beyond the scope of this Guide.

Teachers involved in the subject, who have little or no experience and wish for guidance, may build up a series of case studies at different levels of complexity. Thus, for undergraduate students one can start with the location of material for an essay on culture and simply use an encyclopedia of social sciences. This can then be built up into a more complex search and finally a specific in-depth search can be carried out.

A prominent feature of the social sciences is terminology, and it is desirable to embody in the teaching some material on the construction of index terms, as well as search terms, and to make comparisons. In short, a librarianship approach is more suitable.

Humanities and Arts. Here the whole approach needs to be so different as to put the subject beyond the scope of this Guide. The bibliographical tools have different levels of usefulness. The problem of identification and location of bibliographies requires much more emphasis.

The identification of search features can be taught by a series of exercises at suitable intervals. Some exercises can be based on disparities between features associated with documents (indexing) and features associated with searching for information. The problem is complicated by the words (descriptions or keywords) used to describe the features.

If this idealized instruction can be given before entrance to universities, the suggested instruction to undergraduates will become an extension of their earlier teaching. The students will now be specializing in particular fields of study and in the introductory sessions will consider the occurrence and identification of information needs. The practical techniques will be emphasized, so that students of science and technology can study information retrieval on a scientific basis.

Instruction to Postgraduates. Instruction for postgraduate students needs to be different from instruction for undergraduates because they have received more education, tend to be of higher academic ability, are usually in smaller groups and are expected to spend a higher proportion of time on research and less time in lectures. In addition, there is an increasing probability that they have already received some instruction in information retrieval.

The Guide suggests two alternative approaches: to modify the undergraduate-level material, and to base the teaching on a topic or project.

Modifying the undergraduate-level material is not difficult, especially if the postgraduates have not received earlier instruction. The modifications should, in practice, be based on circumstances. Small groups permit participation, discussion, supervised practical work and individual attention. Their level of ability may permit easy development of the basic material and inclusion of the more complex material; and their experience may be sufficient for illustrations to be drawn from it. One programme that has worked satisfactorily is an alternation of classroom and practical sessions, starting with revision of undergraduate material and followed by a practical session based on undergraduate questions but modified to cover more complex realities. An example of one possible question is shown in Figure 6.3.1. The next classroom session should start with a debriefing of the practical work and should then follow on, as appropriate.

The topic or project-based approach is a substantially different arrangement of the content. It can be useful with some postgraduate groups and most groups of practitioners. The approach is described in detail in Chapter 9.

Instruction to Practising Scientists and Technologists. Instruction for practising scientists and technologists needs to be different from that for students, because their attitudes and backgrounds can be different. It is usually best to reassess the objectives of instruction and how success can be measured. The programme of instruction can then be planned accordingly.

The main difference is that the lecture-room is not a commonplace thing to all practising scientists and engineers, and a few restless figures among them can be disturbing. Engineers tend to be very practically orientated, maintenance and service engineers more so than design or planning engineers. Other differences are a wider scatter of background (different subjects, employers, ages, and so on), and unfamiliarity - they may be unknown to each other and the library used for the exercises may be unfamiliar. The instruction usually has to be given intensively, perhaps over one or more days, rather than extensively over several weeks.

From this it would seem that the theory and principles should be presented fairly briefly and attractively. The emphasis can thus be on practical aspects and particularly practical work.

6.3.2 Differences in Teaching Approach

The basic outlines have been drawn from material developed and used over a number of years and have been influenced by the views and work of a number of different people. They can be adapted for a wide range of situations. However, particular situations or beliefs may require approaches beyond the scope of these outlines. In addition, the teaching of information retrieval to students is still developing, so it is still important to study a wide range of possible variations.

Practical Viewpoint. The viewpoint which has been adopted in this Guide is a practical viewpoint with accepted limitations. The principal limitation is the amount of time available for instruction and the limited number of teaching staff and copies of bibliographical tools.

Experience has shown that a supervised practical session is satisfactory when the group is limited in number and by the teacher/student ratio. Otherwise unsupervised exercises seem to be more appropriate.

If more time is available than assumed in the basic outlines, it is suggested that this should be used for practical exercises, more case studies, or both. The practical exercises can with advantage be graded exercises, and they should be on basic search techniques rather than on specialized information needs. If more case studies are used, they need to be carefully prepared according to the aims of the teacher. A second case study needs to be more than a repetition of an earlier one, but will vary according to what the teacher is trying to achieve.

Theoretical Viewpoint. There is a temptation when teaching practical skills to teach only the procedure to be followed. This is a limited approach, and it is argued that, to learn a skill properly, a student should know why he ought to do something as well as what to do. Thus he can acquire an understanding which can allow him to follow the unusual case.

DESIGN DEPARTMENT | LITERATURE SEARCHING

PROJECT 724 Design of a galvanized steel dustbin for dustless emptying

1. The design must obviously conform to legal requirements. Provide an introductory text on the law of public cleansing.

2. Can you provide diagrams of suggested designs?

3. With wet refuse, corrosion is likely to be a problem. Trace a recent study of the limits of anti-corrosion protection for galvanized steels.

4. Industry faces enormous problems concerned with waste handling. Can you trace an article describing some of the means employed?

5. Is there a research organization operating in this field?

6. Please provide a list of manufacturers of continuously galvanized steel sheet and strip.

Figure 6.3.1: Practical Sheet for Project Work

There are too many variable factors in literature searching for it to be taught only as a series of steps. This was illustrated in a recent first-year undergraduate examination, in which a question asked what procedures would be followed to find information on sewage treatment in France. Specification of a French element demanded an understanding of search techniques, which needed to be reflected in the answer. Even the students who had memorized handout notes, flowcharts, titles of bibliographical tools, and so on, felt obliged to think about the implications of the French element. This example shows that understanding 'why' is important. Some teaching of the theory of searching is implied.

It has earlier been suggested that, if more time is available than is assumed in the basic outlines, this is rightly spent at present, on practical exercises. In the longer term, some of the time may appropriately be spent on practical search techniques, but more of it may with advantage be spent on the theory of search techniques. To do this, one needs to understand the theory better. The work needed to acquire this understanding will be difficult because of the complexity introduced by the number of variable factors.

Librarianship Approach. For the most part, the term 'information retrieval' has been used in this Guide to mean the acquisition of information by consulting appropriate 'places' (shelves or minds) in which it may be stored. The inclusion or exclusion of information from observation (including experiments) as part of information retrieval has been deliberately vague. This reflects the real situation, in which observation must be accepted as a source of information although the necessary skills are usually taught separately from information retrieval. From the librarianship viewpoint, information retrieval is often seen to be more specifically of library use.

This distinction must be clear in the teacher's mind. It is suggested that some scientists, and technologists in general, are objective-minded and often seek an understanding or development of a single idea or concept. Thus they tend to develop 'convergent thinking', and information retrieval seen as the acquisition of information suits this style of thought. By comparison, other scientists and students of the arts tend to develop 'divergent thinking' because they seek to understand the interplay of various concepts, and information retrieval seen as library use may be more appropriate. These are generalizations and must be accepted as such, but if the teacher basically agrees with them then the 'librarianship viewpoint' has a proper place in instruction of information retrieval. However, there is another important distinction between acquisition of information and the use of the library, and this is emphasis.

In the 'acquisition of information' concept the emphasis is placed on the occurrence and identification of information needs and on satisfying at least the key needs. The library as a source of information must be seen in relation to the other sources of information. In the 'use of the library' concept, the emphasis is placed on the contents and the ability to make use of all the stock and services (including inter-library loans). The librarianship approach is ideally suited to students of history, literature, and comparable subjects.

In fact, there is an interface between libraries/information centres and users. On one side there are needs, on the other, resources. The interface is an important part because it is the site of the two viewpoints and there are basic differences between the two sides.

A compromise between the two approaches may be needed for social scientists, who make substantial use of certain categories of material, such as official publications. Indeed, each of the two approaches should accommodate certain aspects of the other. To create a gap between them would be as false as the division between the sciences and the arts.

Some information-retrieval teachers may prefer the librarianship approach; they ought to examine their motives, because their own background naturally favours this approach. The criterion for decision must be the approach favoured by the students.

Finally, if a librarianship approach is being used for scientists and technologists, whose attitudes may be different from those of other groups, an important question should be asked by the teacher after the instruction: "How will the students know which categories of literature they need to search and in what order?"

Information-Science Approach. In reality, a viewpoint which favours a librarianship approach only slightly more than in the basic outlines is difficult to act on. It is easier to adopt one approach or the other outright. If the teacher prefers a middle course, he can use an information-science approach. In essence, this considers information handling from both sides - the sorting and indexing of material and its retrieval. A particular advantage of this approach is that information handling does embody both aspects and students who understand something of the difficulties of information handling, particularly indexing, can possibly get a better insight into difficulties of retrieval. Definition of parameters, both indexing terms and search terms, is thus the crux of this approach.

If this approach is used, it is suggested that it should supplement the appropriate basic outline rather than replace particular aspects. Thus, additional time should be made available. If the approach is used on a trial basis, a small group should be used, preferably at postgraduate or finalist level. The understanding gained is more appropriate to advanced-level instruction and the technique is more suited to those familiar with academic learning. Established engineers, particularly those involved with production, might regard it as 'playing about'.

A typical session with a small group of students may involve a number of items for indexing and a number of search tests. A small information-retrieval system can be produced and searches made. Recall and precision figures can be calculated and a subsequent discussion of the results can be useful. The details can be as the teacher sees appropriate. A distinction should be clearly made between a system which retrieves references to the required information and one which retrieves the required information itself.

Having made this distinction it may be fruitful to consider modification of the input to determine the effect on potential output. Thus hierarchical arrangements and controlled language can be used.[1]

Finally, a brief description may be given of some of the recent research in information science.

Theoretically Orientated Groups. The advanced level of study followed in the information-science approach can attract the interest of groups with a common background, notably mathematics, computer studies and very theoretically-orientated groups. For the most part, such groups can be taught in the same way as postgraduates provided diplomacy is used. For example, discussions, workshops or seminars may be more appropriate than lectures, instruction or teach-ins. The group may even be librarians, either postgraduate students or other members of this increasingly extensive profession.

The content is likely to be based on the information-science approach used in the previous section but extended to include work on automatic indexing and retrieval of texts and similar topics. These circumstances are likely to be unusual and any detailed notes will soon be outdated. Certainly the teacher needs to use retrieval skills himself and must have access to adequate library collections. He is likely to aim at being informative rather than instructive.

Extensive Approach. For the most part this Guide is intended for realistic situations which exist at the date of publication. In general, it is advocated that instruction should start with children and continue through to in-service teaching. If this approach is adopted widely, it will need to be modified at the undergraduate and postgraduate levels of education. One more immediate possibility is that time will be more generously allocated and can be used, without overburdening teachers, for an extension of the programme of instruction or for inclusion of the subject at more than one stage of an undergraduate's training. Given a choice, the latter use is to be preferred because, it is believed, extending instruction over an educational period is for most people more important than advanced instruction for a more select group.

In using the extra time initial attention can be given to practical search techniques and associated practical exercises. In the ideal situation one should start by assessing existing information-retrieval skills and setting tasks slightly more difficult than existing skills in an environment where users can find the answers and develop their skills. Exercises can be graded in a formal manner and based on the potential information needs which the individual ought to be able to satisfy himself. An alternative is to use more informal teaching techniques such as 'in-tray' exercises, simulation games and role-playing. These require careful planning, and on paper appear to be remarkably attractive. The results can be disappointing, however, basically because they are dependent on chance. An incomplete group for role-playing or an unresponsive individual can mar the session. However, more experience may give clear guidelines for such sessions. In an ideal situation a small room within the library should be used and fitted with the necessary equipment. The graded exercises are used simply to develop information-retrieval skills and the simulation and workshop sessions to develop skills in identifying information needs and in following the relevant decision processes (sources, time, probabilities).

---

[1]

Source material on this theme can be found in FOSKETT A.C., The Subject Approach to Information . 2nd edition. London, Bingley. 1971. 429p.

In the long term a library workshop can be developed, if simulation proves valuable. Self-instructional material can be made available within a supervised area, samples of various files can be produced, and the room can become the base for the practical exercises within the library.

Having thus extended the development of basic search skills, the second stage is to provide more theoretical knowledge of information retrieval by studying the effects of variables, including variations in indexing policies. (See 'theoretical viewpoint' section above.) After that, the next expansion is a more specialized treatment, as considered in the next chapter.

Intensive Approach. The intensive approach has been catered for in the basic outlines for short courses given in the appendices.

In large intensive courses lasting more than a day consideration can be given to evaluating the results of a literature search. This aspect of information retrieval is perhaps best studied in detail when considering the retrieval of data and statistics.

Use of digested information can be studied in short intensive courses, particularly those for induction or people returning to education. For secondary-school pupils and new university students the advantages of authoritative encyclopedias can be described, or, better still, demonstrated. For scientists and engineers review articles are important and should be mentioned at some stage. For social scientists, Keesing's Contemporary Archives, with its relatively succinct news items, can be a time-saver, provided that one accepts that it contains only the essential detail.

6.3.3 Differences Arising from Locale

As in the last chapter, much of what has been described applies to both developing and developed countries. Thus, little can be said of particular circumstances in developing countries.

The basic search technique may apply to a higher proportion of the total literature in those countries which have a restricted range of bibliographical tools. Thus, instruction in the basic technique may have to extend beyond the period indicated in the basic outline.

6.4 POSSIBLE DEVELOPMENTS

It has been argued in this Guide that a mixture of teaching methods can be used to overcome the limitations of particular methods. It may be that a particular method has been overlooked, but it seems over-optimistic to believe that any new teaching method will supersede existing teaching methods for basic search techniques.

More likely, there will be improvements in bibliographical tools. They are unlikely to supersede instruction in information retrieval, but they could bring changes in the detail of teaching practical search techniques. For example, if they make it easy to locate physical, chemical and engineering data, the retrieval of such data could become the basic search technique for scientists and engineers. Similarly, statistics could be used for the basic search technique for social scientists, especially economists. There are certain advantages in using data as the basic search technique. For example, practising engineers recognize their need for data more easily than their need for reading material to build up knowledge. However, to use data retrieval as the basic technique requires the availability of adequate bibliographical tools, including, if necessary, a subject index to reference books.

New bibliographical tools are unlikely to necessitate changes in the near future. New tools such as citation indexes have appeared, and have added greatly to the range of available tools, but the pattern of teaching has accommodated them relatively easily. A greater problem is the cost of these new tools.

6.5 CONCLUSION

Given sufficient attention to the theoretical basis, practical search techniques are very important. It is difficult and probably unsatisfactory to teach basic practical techniques by a single teaching method and a mixture of methods is advocated, including practical exercises. If the instruction is to be of value for some time, well-prepared notes should be handed out. Part of the approach should illustrate the decisions and variable factors involved even in basic search techniques. All this can lead to an indigestible amount of information which can be made more presentable by the use of case studies and teaching aids, particularly audio-visual material.

The success of the teacher in developing a basic search technique should be measured by the effect on the students' study technique and work performance, but this is impossible to all intents and purposes. Examination questions based on the basic search technique have shown that many students have been able to grasp the principle of basic search techniques, and it is interesting that answers have varied in their style of presentation, ranging across flowcharts, numbered steps and interpretations of the various teaching methods.

6.6 CHECKLIST USING THE APPROACH GIVEN IN THE GUIDE

1. Have you ascertained the main aspects of the method of finding information which will be taught by you?

   The aspects can be:

   N.B. Modification will be according to personal beliefs.

2. Have you detailed these main aspects into practical useful information needed for developing practical skills?

   The details can be:

   a. Search tools - book-lists,
                   - abstracting and indexing journals,
                   - reference books,
                   - bibliographies and bibliographical guides.

   Choosing the appropriate source is the key factor.

   b. Search strategy - adopting the principles of problem solving is advised,
                        together with lists and notes for the novice.

   c. Search terms   - appreciation of the difficulties,
                       - possible courses of action.

   N.B. Modification will be according to local circumstances and choice of alternative approaches.

3. Will a user who understands your **proposed** content be able to satisfy the educational objectives you set in the preparation for the course?

   Review your content and your objectives.

4. Have you used sufficient different working methods to deal with different user **abilities** and to meet both immediate and future needs for guidance on information retrieval?

   Use the following as appropriate:

   Counselling - flexible, effective, timely, takes time, student initiated, specific.

   Practical exercises - highly desirable (essential even), based on actual processes linked to educational objectives.

   Case studies - illustrative, reinforced ideas and concepts, student participation can be useful or lack response.

   Printed guides or notes - semi-permanent record for students.

   Flowcharts - if good, can be very useful.

5. Have you simplified the real situation? What additional material and guidance is needed by the users? How are you going to follow up this lecture or this part of your programme? See Chapter 7.

6. What reactions have you had from your group? How does it compare with other groups? What are the causes of the differences?

   Revise your ideas and notes, if necessary.

7. Course content 3—Development of search techniques

Basic search techniques can be developed along two principal lines: first, repetition and variation of the basic search technique and second, extending the coverage to include specialized categories of information. It is advisable for the teacher to include some development along both lines. The relative priorities and emphasis will depend on a mixture of circumstances and personal beliefs.

An approach to development of search techniques that works well for students is to concentrate on specialized aspects in the formal teaching (lectures and question-based practical work), and, at or about the same time, have the teaching departments set students projects involving a high proportion of basic information-retrieval skills. This approach has many advantages in terms of motivation, grasp of the whole topic, individual teaching, participation and general co-operation. Similarly, instruction in information retrieval of physical data can be correlated with practical laboratory work in the syllabus. The specialized aspects can be covered in a lecture with supporting practical work. A typical outline for a third lecture might be:

Specialized information needs

(detailed according to the background of the student group)

Current awareness	– selective dissemination of information, – list of recent articles, – scanning.
At the desk/in the field	– 'desk-top libraries', – personal files, – information services.
Developments in libraries	– inverted files, – citation indexes, – computers.
Finale	

NOTES:

1. This lecture outline is based on instruction of students. For practitioners, it is preferable to re-orient the whole instruction so that routine information needs, such as data, take precedence. A suitable programme is given in Figure 3.2. The content for any selected aspect is the same.
2. As in earlier chapters, it is intended that the tutor should select material for inclusion in lectures and practical work rather than include all possible topics or even those detailed in this chapter. The needs of students or practitioners should be the main deciding factor and the immediate needs should have priority. Selection of material for practitioners can be in response to interests expressed by specific groups.
3. The range and depth of coverage need substantial pruning for first-year students. The actual content is often more appropriate for final year undergraduates, postgraduates and practitioners. In other words, the target group for the material in this chapter has been one of mature, well-educated and intelligent students such as postgraduates or an exceptional final year undergraduate group.
4. It is advisable to keep a checklist of specialized information needs and to itemize those that are to be discussed.

7.1 SPECIALIZED INFORMATION NEEDS

For scientists and engineers, the most important specialized information need is data and factual information. Consequently, the amount of material and time devoted to it should be proportionally high. The same is true of statistical information for economists and managers.

In theory, data and statistics are very similar but in practice they are reached by different means of access in the literature.

7.1.1 Data and Factual Information

NOTES: The term 'data' is here intended to mean the measured values of physical, chemical and engineering characteristics of phenomena. Unfortunately, there is confusion in its use. The term 'data input' has been used to describe information input to computers for storage or handling according to input of programmed instructions. Computers can be used for mechanized information handling and retrieval, and the output may be 'references' in the form of bibliographical details of documents which can be subsequently consulted. If instead of references the output is actual information, this is sometimes referred to as 'data retrieval' and can be physical, chemical or engineering data. We have used this 'data retrieval' (as distinct from reference retrieval) for these **data**, including statistics provided by mechanical searching. When the output is information which can be read and digested, we have used the terms 'text retrieval', 'part-text retrieval' and 'abstract retrieval'. The practice of other people is often different.

Data and factual information can be considered at a number of different levels. <u>Basic Level - General</u> (Suitable for non-scientific groups and subjects; coverage in secondary education before science specialization is developed; general public; general introductions).

Real-life tasks, unless repetitive and devoid of intellectual effort, create needs for facts which are often mundane and include addresses, time-table information and so on. The needs can often be classified as follows:

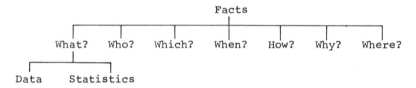

Figure 7.1.1: Real-Life Situation Issues

Some of these categories are very important in certain fields; for example, 'what' in science, technology, social sciences and management; 'when' (dates) in history; 'where' in geography and 'how' in cookery, and so on. Some types of bibliographical tool also specialize; for example gazetteers cover 'where'. If one goes into these aspects in depth, searching can be very advanced, requiring skill and time. In addition, factual information at an advanced level requires interpretation: the scientist compares and evaluates a new experimental result with associated earlier values; the historian compares a new 'find' with other associated records. Scientists, historians, physicians, lawyers and others are aware that incorrect facts may be obtained by reading something in a book. The rule is always check:

Is this fact likely to be correct?)	Faults in determining, selecting,
Where did it come from?)	presenting, and printing (responsibility
How was it known?)	of author, editor and publisher).

Have I understood it correctly?)	
Were limits specified?)	Faults in digesting information
What units were used?)	(responsibility also with reader).

Sometimes invalid 'facts' reveal themselves harmlessly, even if frustration results; for example, a change of address or telephone number of a person or company. On some occasions the use of invalid 'facts' can be very serious, particularly in science and technology. This is a subject in itself and can be given more detailed attention if time allows and interest is shown.

The practical search techniques which apply to state-of-the-art searching also apply to factual information:

Search tools,
Search strategy,
Search terms.

Search tools. Reference books are significant and need particular attention. Familiarity with particular reference collections is invaluable, but even the newcomer to information retrieval can acquire a degree of self-sufficiency by using commonsense. When readers cannot find information, they should consult librarians, who should have a good knowledge of the reference collection and other sources of factual information. The time concerned is mainly search time (excluding time spent on evaluating facts), because the amounts of information are relatively small.

Search strategy. The strategy is clearly similar to the basic search strategy. The first part of that strategy relating to reference books is particularly important.

Search terms. The choice of search terms is based on the same findings as for the basic search strategy except that the quality of indexing in reference books and books is generally inferior to that in abstracting and indexing journals. There is a greater need to check possible headings because cross-references do not as frequently guide the unwary to the right term. Many professionals (librarians and information officers) often check the contents list of a reference book as well as the index. They also make unkind comments about reference books without indexes.

Searching for factual information can frequently be more difficult than searching for references. This is because the possible search tools are more numerous and the choice of search terms is more demanding. On the other hand, successful searching need not be so comprehensive because the acceptability of located material can be based on smaller amounts of information. Similarly, evaluation is often easier and more obvious.

The preceding paragraphs indicate a possible content for course material at this level. The presentation may be based on discussions or question-and-answer sessions or simply on a written lecture. The former are probably better provided that the objective is (like the content of this section of the Guide) to extend understanding of basic search techniques to factual information.

Basic scientific level. (Suitable for students specializing in science or technology, particularly in the first year at university or college).

The main objective at this stage is to extend the actual information-retrieval skills, so that students can find items of physical, chemical and engineering data which they may require as part of their studies. Ideally, the skills should be specified in precise terms, such as the ability to compare the experimental results of their laboratory work with those in the published literature and to explain possible causes of differences. Such an objective goes beyond information-retrieval skills but it clearly involves the student's academic attainment and is one aspect of coordinating instruction with the teaching department.

A secondary objective is that of the previous section - to extend understanding of basic search techniques by treating data and factual information as a variation.

It is possible to achieve a balance of the primary and secondary objectives. One method is to use a handout which contains detailed guide-lines on finding information (see Appendix 2) supported by practical exercises to develop skills. The understanding can be developed by 'running-through' the handout during the lecture and then discussing the variations from the basic search techniques - tools, strategy and terms. Reference should be made to invalid data in the handout and discussion, or possibly in the briefing for practical exercises.

With generous allocation of time, i.e. more than three hours for lectures, one can devise a class exercise involving a single item of data taken from several reputable reference books. The choice of item should be made so as to include misrepresentation, misprinting and misreading of data.

The topic can come under a general heading of evaluating and using information, and can be developed so as to promote consideration of the penalties of not seeking and evaluating published data.

Higher level. (Suitable for scientists and technologists who understand the basic level of searching for data and have a need or wish for more depth - especially postgraduates and practising scientists and engineers.)

The whole of the basic level of instruction can be expanded to cover the evaluation of published data and alternative sources of information. This greater detail is particularly desirable for mathematicians, physicists, chemists and design engineers. It is also desirable that the teacher should have had some scientific or technical education.

Published data can vary according to:

 1. accuracy,
 2. reliability,
 3. authoritative evaluation.

Accuracy for the scientist is mainly a matter of reducing experimental errors to a minimum. He places emphasis on the repeatability of measurements. With certain items such as the speed of light the degree of accuracy is known. For technologists using engineering data, a degree of variation can be anticipated from minor constituent factors. For example, the strength of a specified steel will vary according to grain size, slight variations in compositions, microscopic defects, surface finish and specimen type. In such a case, working data can consist of typical values or minimum specified values. If a person is concerned with accuracy of data then there is a need to know the limits of the values, and this may mean authoritative evaluation or reading details of the original experimental work.

Reliability is concerned with data (and other information) being what it is said to be. Information can be unreliable because of misprints, misrepresentation, misreading or misinterpretation (see previous sub-section). The obvious precaution against unreliable data is to compare figures from two or more sources in the manner needed to improve accuracy. Often the scientist or technologist does not want very accurate information but is satisfied with an approximate value; this means that he does not want to spend the time needed to check accuracy. Fortunately unreliable data can often be detected by simply checking that a value appears correct. This useful habit saves time and frustration in accordance with the rule "measure twice and cut once".

Conscious checking that a value appears plausible is a matter of awareness, and can first be done by looking at associated values, if the material is in tabular form. This is useful with speeds and feeds for machine tools, conversion tables and similar items. A second check is to visualize the data in their component aspects; this can be useful for working loads and stresses. All this sounds tedious but in practice can become an almost sub-conscious activity.

The ideal way of minimizing difficulties with inaccurate and unreliable data is for central authoritative organizations to make up-dated evaluation available in a suitable form.

Overdesigning (or guesswork with a margin for error) is an apparently simple and quick alternative to seeking data. The result of such a practice, even in simple cases, can be unnecessary expense. In the case of a single purpose-built design or system this can be less-than-optimum performance. For multiple products and systems, the cost accumulates and can soon exceed the cost of seeking accurate data. It is possible by overdesigning to overlook the predictability of a product or system, and the consequences of this can be serious. It may be desirable to have certain components fail at a predetermined value so that the remainder of the system is safeguarded.

Instruction of higher-level users on the validity/invalidity of data should be by involvement and discussion, during which the teacher aims to stimulate members of the group to consider the factors and precautions and then contribute constructive ideas to the debate. Stimulus can be introduced by case histories and examples. One source of unreliability can be vague general statements. For example, a storage tank said to have held a 'paraffin-type' chemical had in fact contained an uncommon substance similar to paraffin but with a much lower flash point. An explosion during dismantling killed several people.

The second part of the higher level of instruction is easier for the teacher. This concerns seeking data beyond reference books. Even at the basic level, abstracting journals can be discussed and reference can be made to specific abstracting journals such as Chemical Abstracts which have proved useful in locating data. The unpublished sources (reports) and direct contact with organizations can be then considered. Data can also be linked to products and standards and information on these may be considered separately.

7.1.2 Statistical Information

Statistics like data are a particular form of factual information.

Most scientists and engineers need a relatively brief treatment of statistics, particularly since adequate guides to the literature exist. To develop the subject beyond these guides will take more time than is normally available. A possible exception is engineers interested in various aspects of management. Statistical information is considered in detail for economists and social scientists in Appendix 3B.

Statistics can be regarded as the measured values of a particular characteristics defined geographically and by date (data may be defined by time and space). National bodies have collected and collated statistical information for the benefit of society. International bodies have similarly produced collations from various sources so that comparisons can be made. The resulting bibliographical tools are useful for general, and even some more specific, statistical needs. For other specific needs, there are numerous statistical compilations which may prove useful and also a number of guides to sources of statistics, which can aid location. Problems still exist, mainly in establishing the breakdown of figures into different facets. Thus, the user who is interested in the purchase of footwear may find that one source gives a breakdown by income level and another by age, whereas he may desire breakdown by occupation.

7.1.3 Product Information

Studies of engineers' information needs have shown that the most frequent need of design engineers is for information on available products and materials. It is believed that some other categories of engineers have very high demands for product information, though engineers as a whole tend to qualify and quantify their information needs by what is available rather than by what is possible. The need is possibly less significant for scientists; but it still exists, particularly for users of chemicals, unusual materials and experimental apparatus. (See Appendix 3A)

Access to product information is rather unbalanced in many countries. Manufacturers and distributors of products and materials produce carefully thought-out and often expensive catalogues and brochures, which they try to distribute to their best advantage. At the end of the communication link an engineering designer is attempting to procure particular components or materials which meet certain standards and specifications and which make a compromise between cost, size, reliability and performance. What the engineer would like ideally is an easily accessible and relatively comprehensive manual which indicates the range of items available from different manufacturers and suppliers. Some people think that this apparent imbalance can be redressed if a portion of the money spent on advertising and publicity is spent on producing services for design engineers. Some organizations have attempted to do this, with the result that some Services have established themselves quite well and others have been withdrawn because of expense or of underestimation of the problems. Comparisons are not easy and some manufacturers feel that evaluations do not accurately assess the virtues of their products in comparison with others, especially when comparable products are improved over different time scales.

While the current situation is not as satisfactory as some people would wish, it must be accepted and changes should be made when seen to be beneficial. Systems for providing comprehensive information do exist and are being developed further. Organizations which wish to purchase these systems can do so. If not, the wide range of catalogues can serve most engineers to some extent and, although costly for the supplier, can bring him supplementary benefits. In addition to publications technical sales representatives are regarded as valuable sources of information and as useful communication links between suppliers and customers.

7.1.4 Standards Information

Many of the products, materials, processes, and procedures used in industry are made to specifications given in various standards. These standards are the scope and limitations considered acceptable by authoritative bodies. The International Standards Organisation (ISO) provides world standards, many of which have been adopted by various national bodies. Many nations have an official authority which also establishes standards either by adopting a relevant ISO standard or as a result of the deliberations by panels of expert and interested parties, including manufacturers. Usually they produce a yearbook (sometimes called handbook) which gives brief details of each standard and serves as an index. Ideally, the organization will also collect standards and yearbooks of other bodies and serve as an information source on all aspects of standards. If these ideals are met, the acquisition of standards information is often straightforward.

Standards specify limits, and minimum values in particular, rather than ideal or even typical values. In practice, it can be important to realize the distinction between minimum values and typical values. Standard specifications are a source of accepted limits and not necessarily a source of data. However, they serve a very useful purpose merely by giving users these known and accepted limits for a wide range of relevant products, processes, and procedures. They save him from being subjected to a significant new set of conditions with every new batch of components, semi-finished goods, or materials. The national authority or its financial supporters should have an interest in providing information on unusual standards because these are normally associated with export orders, which in turn are often encouraged by national authorities.

In drawing a plan for a particular objective then an individual engineer accommodates standard specifications according to his vocational training and experience. The relevant standards are normally detailed, so that the appropriate document can be consulted by anyone involved in the project. A standard is normally indicated by the initials associated with the authority, followed by the number and part - where appropriate - of the relevant document and possibly the date. A particular standard specification may refer to a range of possible choices - the classic example is materials specifications, especially for steels. These choices may also be referred to by a sub-series of numbers and letter prefixes. In cases where such a sub-series exists there is sometimes a tendency to refer to just the subseries coding. A typical example of an entirely accepted series of standards is produced by the German Standards Organization (Deutsches Institut für Normung - DIN); one of the steels series is DIN 17100; a particular steel is St 33.2.

In the teaching of standards information to engineers, the teacher is advised to describe the value of standards in one or two sentences, to show the form of typical standards used in the home country, and then to consider the occurrence and identification of standards information needs. The occurrence is any product, material, process, or procedure which should follow accepted standards and for which appropriate standard specifications are available in the country. The rest of the teaching should deal with searching. A standard specification on a plan or contract can usually be recognized by its letters and numbers (possibly with part and date). Location of standards so quoted can be traced through the yearbook (or handbook) of the issuing body. Cases of difficulty can be referred to the national authority on standards in an ideal situation. Where bodies have limited ability to provide information one must resort to the other informative bodies, such as professional institutions. It is useful to add titles of standard specifications because misprints of numbers can occur and confusion may result.

Although it is believed that location of relevant standards is straightforward provided care is taken, experience shows that errors can occur. Typical causes of error are due to the closely related coverage of more than one standard and the limited indexing involved in producing subject indexes for standards. In practice, three other causes of difficulty may arise: one is the practical reference to a standard, typically the sub-series for steels; the second is the involvement of unfamiliar series of standards; finally there is the attempt to use items produced to one set of standards to meet the specifications of another set of standards.

One should perhaps mention that there are also standards concerned with bibliographic topics, for example the British Standards Institution's BS1000 on the Universal Decimal Classification and those supplying glossaries of terms used in industrial processes or management.

7.1.5 Materials Information

Sub-headings for specialized types of information cannot unfortunately be mutually exclusive. Materials are often regarded as semi-finished products and can be located through product-information retrieval systems. Similarly, they are often specified by standards and details can be retrieved through standards-information retrieval techniques. In addition, particular techniques can be used to acquire information on materials which have not been covered by the two previous headings, mainly because some bibliographical tools exist which deal specifically with materials.[1] A select few are worth particular mention.

7.1.6 Patents Information

Each country normally has an organization handling its patents. The intention is to encourage people to disclose details of their inventions in return for limited legal rights in their exploitation. In practice inventors, on payment of the appropriate fees and on giving careful attention to the detailed procedure, can restrict the manufacture and sale of their inventions in the relevant countries provided they are accepted as patentable. The result can be disappointing; or it can bring financial benefits from licence or manufacture.

Patents and patenting form a complex subject, generally outside the scope of this Guide. Any doubts about complexity can be dispelled by opening any relevant legal book on patents.

Patent specifications are a source of technical information and, indirectly, a source of commercial information. Problems and difficulties do arise however - different practices in different countries, interest stretching beyond national boundaries, and orientation or bias in classification schemes and indexes for patents. A classification scheme can reflect an intent to classify by the principle of the invention rather than by its application. Similarly an indexing scheme may be so precise that the familiar object projector is termed a projecting lantern and the term projector is used for something far less common.

---

[1] For general data information:
   TOULOUKIAN, Y.S. (ed.): Thermophysical properties research literature retrieval guide. 2nd edition. (3 volumes). New York, Plenum Press. 1967
For medicaments and pharmaceuticals:
   The Merck index: an encyclopedia of chemicals and drugs. 8th edition. Rahway, N.J., Merck. 1968. 1713p
   Martindale, W.: The extra pharmacopoeia. 26th edition. London, Pharmaceutical Press, 1972. 2320p
For steels:
   Stahlschlüssel. 8th edition. Murbach (Fed.Rep.of Germany), Verlag Stahlschlüssel. 1968.
For metals in general:
   ROSS, R.B.: Metallic materials specification handbook. 2nd edition. London, Spon. 1972. 833p
   WOLDMAN, N.E. (ed.): Engineering alloys. 5th edition. New York, Van Nostrand-Reinhold. 1973. 1427p

It is advisable to teach the detail of obtaining technical information from patent specification only to those groups where it is necessary. Such groups will be relatively advanced students in appropriate subject fields and in circumstances where sufficient time and resources are available to do justice to the subject. For this reason, the subject is regarded as very specialized, even in the area of specialized information needs. Some tutors will find the need to compromise completely ignoring information from patents and covering it in detail. One compromise can be to include any suitable publication produced by a relevant official organization in reasonable depth.

7.1.7 Legal and Medical Information

In a stable society the legal procedure is dependent on information retrieval. Indeed, lawyers, and to a great extent physicians and patent agents, are specialists in information retrieval. In all three cases qualification is a necessary preliminary to the practice, and information officers in general must recognize this. Specialist law and medical libraries often exist as separate units.

The two areas, law and medicine, are very clearly defined and relate to individuals and their personal interests. The use of bibliographical tools in the areas and the understanding of located material is dependent on a core of subject knowledge and on specialist terminology. Familiarity with, and regular use of, the bibliographical tools is also beneficial.

For the teacher and practitioner of information retrieval it is necessary to recognize and accept the specialist nature of the fields. Advice must be either bibliographical or referral and interpretation of the literature must be left to those qualified to do so.

Because of the specialist nature of information retrieval in legal and medical fields, they can prove interesting areas of study for the student of information retrieval because they may have implications for other fields. Legal information is probably the better choice for it is more clearly defined, better established, and it is possible that theory and practice are more closely associated than in the medical field.

As an example of interesting study in the legal field, the incidence of lack of precedence is clearly recognized and significant new developments are clearly identified and documented. This feature seems to be very desirable but is difficult to incorporate in other fields, including medicine.

There is an overlap of interests between these two fields and others. A simple case is patents, in which the procedures are legal but the content can have scientific and technical value. Again, management students need some education that enables them to operate within the legal limits of society. It is suggested that some instruction for management students should be from legal specialists and should include the retrieval of legal information.

In the medical field, some aspects of information retrieval may leave something to be desired, but in one respect - current awareness of important new developments - information flow is remarkably efficient. In part this is due to the responsible and authoritative role of certain journals, but the public also are rapidly informed of some events.

7.1.8 News Items

Some significant events have news value and are reported as such. Although news items are primarily intended for current awareness of topical events they can have a more permanent value.

News items are included in some periodicals as well as newspapers. Specialist bibliographic tools have developed to cover them.

When searching retrospectively for a news item, the date of the event and its geographical importance are significant. If the event is of international or national importance, the national newspapers will have included coverage. Some of the key national newspapers have published indexes, e.g. the United Kingdom's Index to the Times. If the event is of interest to only the immediate locality, it may be necessary to consult the provincial newspapers. Back runs of these are generally available in public libraries and sometimes can be consulted on the publisher's premises. Some newspaper publishers maintain their own indexes, which they will search on behalf of an enquirer. Many national libraries also serve as archival stores for newspapers.

Other bibliographical tools are worth noting. Brief outlines of world events can prove useful for the searcher requiring search material easily and quickly. Keesings Contemporary Archives of World Events serves this purpose very usefully. Research Index is quickly available and deals with commercial and industrial news. The speed of compilation necessitates broad subject indexing, but this 'bibliographical tool' is both unusual and useful.

7.1.9 Commercial Information

This can be considered essentially in two parts: information about companies themselves and information about the state of the market.

A company readily makes available information on its products and services but is naturally reluctant to reveal financial information and future prospects. National practices and traditions usually influence the availability of such information.

It is important for an information officer to emphasize the ethics of this, since there are often 'improper' or illegal sources of information which should clearly not be used.

The logical approach to seeking information on companies is:

1. use of known relevant directories and any available specialized documentation services covering companies;
2. reference to a central register of companies and the deposited company papers (or direct to company itself);
3. reference to further sources of information such as associations, special libraries, information services, newspaper indexes, as appropriate;
4. employment of an expert to seek obscure details and to interpret available information.

Information on the marketing of products is linked to product information, but is more associated with changes in demand and supply. Literature is only one type of information source and much valuable feedback of information is readily available from customers and consumers. The sources of information are varied and overlap considerably with those for other types of information such as news items, patents information, product and statistical information etc.

7.1.10 Information for Education Studies (Theses and Dissertations)

Students, particularly postgraduates, are often expected as part of their education to carry out in-depth studies which are to be written up and subsequently assessed. Some of the work done is significant and access to it is often possible, through bibliographical tools such as <u>Dessertation Abstracts</u>.

7.2 CURRENT-AWARENESS INFORMATION

So far emphasis has been given to retrospective searching for information which arises from an identified need.

Current awareness is important in its own right, however, and must be given sufficient prominence. It is advisable for the teacher to allow adequate time for details even if it means neglecting some of the less specialized information needs. The needs of users change as their courses develop. Undergraduates have the greatest need of information for learning, and therefore books are important to them. Postgraduates may need to carry out detailed retrospective searches. As students advance, they develop more specialized information needs. As soon as they leave university or take up research their need for current awareness becomes strong. When teaching practitioners, current-awareness information is best treated separately - and possibly by a completely different approach.

In theory, keeping up-to-date is little different from 'finding out'. The three main external sources of information still apply. Oral sources, for example, range from formal conferences to informal talks over coffee. Observation is directed to exhibitions, tours and similar visits. For recorded sources, the 'literature base' can be smaller and may be limited to selected publications of the current month.

In using the literature, there are three main methods of keeping aware of current developments:

1. Selective dissemination of information (SDI);
2. Current-awareness bulletins;
3. Scanning current literature.

If current-awareness is important in a particular field or to an individual, it is advisable to use at least two of the three methods; otherwise there are limitations. The preferable combination is selective dissemination and scanning. If the former is not available, a current-awareness bulletin can be substituted.

7.2.1 Selective Dissemination of Information (SDI)

The production of the main abstracting journals is now done by computer type-setting and modified magnetic-tape versions can be used for selective searching. Some magnetic tapes can be purchased and access to selected printout is available in many cases.

A magnetic-tape version of an abstracting service stores in machine-readable form the information contained in the printed abstracting journal. Sometimes additional information is placed on the tape to extend the scope of the searching techniques. These tapes can be searched for a selected combination of elements which appear in particular records and these can be printed out and passed (disseminated) to the user. The elements can be names of authors, titles of periodicals, or subject terms. The latter can be controlled vocabulary terms or words used by the author or indexer in the title or abstract.

A number of factors need to be taken into account. Firstly, the material printed out depends on the quality of the data or literature base and the flexibility permitted in combining the elements. Secondly, the relevance of the output depends on the skill with which the combination (or 'profile') of elements accurately reflects the interests of the user. Work has been done on reorganizing records on the magnetic tapes to cover more than one issue and to subdivide the subject-matter so that searching is made simpler and cheaper. The literature bases covering several months can be used for limited retrospective searches and the differences between current-awareness and retrospective searching can be seen to be a matter of degree.

As to the costs of SDI these still appear to be relatively high and beyond the reach of those countries with a limited scientific population. It has been reported that scientists in isolated situations in Finland appreciate SDI more than those in central locations. Elsewhere it is often necessary to provide the services for a substantial scientific population, and for this purpose international links have been developed.

SDI systems are still under development and for the individual scientist it is necessary to know which services are available and at what price. These can then be compared with the alternative methods of acquiring current awareness and the alternative demands on available financial resources. If SDI is used, the profile needs to be as accurate as possible. Ideally it should be developed jointly over a period of time by the scientist and a trained intermediary, preferably one with subject knowledge.

SDI is normally based on magnetic-tape records and large computers. A minority of information services have attempted to provide comparable SDI by manual means without using expensive equipment. Some developing countries may be tempted to emulate such processes since they are, after all, an expansion of the service provided by a good information officer when passing selected items of interest to individuals. It is suggested that for a few individuals, say up to twenty in number, current awareness is best provided informally based on information officer's knowledge and card files. Beyond this, first consideration should be given to a current-awareness bulletin and then a manual SDI service examined in the light of experience of providing such a bulletin.

7.2.2 Current-Awareness Bulletin

Before magnetic-tape versions of abstracting journals became available, one method of maintaining current awareness was the issue of a bulletin. This is still a viable method for numerous situations, including those in developing countries. The bulletin can be provided in two basic versions. One simply copies the contents pages of recent periodicals and reports, and the other provides bibliographical details (with or without abstracts) in some basic subject grouping.

Of these alternatives many prefer the bulletin which provides references grouped by subject. The advantage of this method is that the individual scientist or engineer has to scan only those sections that interest him and so minimizes the time and tedium of searching.

Copies of contents pages can be provided by photocopying the originals or producing separate lists. In both cases the original may be the copyright of the publisher and permission may be needed. Ready-made bulletins of contents pages are available in a publication called Current Contents produced by the Institute for Scientific Information in the United States. This Institute also supplies one of the magnetic tapes for SDI and obviously believes they are a viable alternative. Since its introduction, Current Contents has undergone several improvements, which implies that current-awareness bulletins have considerable potential.

The third alternative is to circulate the current issues of periodicals themselves. In theory this is an attractive method for both user and librarian, but is not so in the practical experience of many. Hesitant circulation results in stockpiling on unnamed desks and slow arrival in batches at the lower names on the list. This can be avoided to some extent by detailed and time-consuming control by the library staff.

As with the computerized system, there can be a link between current awareness and retrospective searching. For the library to build up a card file of references may be worthwhile in certain situations; it thus creates a bibliographical tool according to the needs of the organization. It can then introduce a current-awareness bulletin, grouped under the same subject terms as the card file.

7.2.3 Scanning

Scanning is browsing through recent acquisitions, periodicals in particular, as a means of keeping aware of developments in science and technology and as a possible means of encouraging serendipity (the happy and fortunate finding of items that are not expected). It is a possible alternative to selective dissemination of information and current-awareness bulletins, it is better used as a supplement.

Scanning is a useful method of keeping up-to-date even when more sophisticated methods are used. It provides for general background and for the occasional item which has not been found by other means because it could not be foreseen.

Some guidelines have been drawn up specifically for scanning by engineers, following a survey of the amount of time engineers believed should be spent on current awareness. It is suggested that between four and ten periodical titles should be scanned, the number depending on the individual, his subject background, and the availability of periodicals. An electronics engineer or aeronautical engineer will probably scan ten periodical titles. The maintenance and plant engineers need only scan four periodical titles. A mechanical engineer may well be in the middle of the range, and for him selected periodical titles should be chosen as follows:

> 1 general science periodical,
> 1 general engineering periodical,
> 2 periodicals for the relevant subject field, and
> 1 or 2 abstracting journals.

In some cases, it may be possible to suggest specific periodical titles. However, recommendations as to periodical titles are better made by subject tutors or at least subject-specialized librarians. Account should be taken of the available knowledge of literature in particular subjects.

The logic of the suggested range of choice is fairly obvious but should be repeated here. The general science periodical is suggested so that general scientific developments are noted. Similarly the engineering periodical can bring to light developments in that area. The two subject-based periodicals are intended to give more specific coverage and finally the abstracting journals are suggested as a source of material which is of interest to the specialist but not necessarily covered in the other journals.

First-year students need not give detailed attention to current awareness; but it is better that they turn to it slowly, as their course proceeds, than abruptly on leaving the university to start work. Current awareness can well start with one periodical in the first year at university.

7.3 AT THE DESK/IN THE FIELD/ON THE SITE

One must accept that there are differences between student and practitioner, even if it is undesirable for them to be as great as at present. The student is primarily learning and acquiring skills and knowledge; he expects study facilities, including libraries. The practitioner expects to use his skills and knowledge to satisfy some of society's needs, and for this he expects payment. He also expects a place of work which is appropriate to the job, and this desk, location, or site may be remote from the facilities that can supplement his skills and knowledge. These facilities can include analytical services, testing equipment, library and information services, and welfare provision. The employer may provide means of overcoming the remoteness of these facilities, but these can be expensive and there is a tendency to provide only those of proven and immediate need within available finances. For the practitioners this may imply that provision is limited and he may have to request certain facilities and attempt some degree of self sufficiency. Library provision can be one of the remote services and attention needs to be given to information acquisition in such remote circumstances. Remoteness can be taken to mean the distance beyond which an individual is not prepared to travel to the library for a particular item. For items of factual information such as telephone numbers and data the distance can be around 200 metres. The result can be the use of more expensive or less satisfactory alternatives. A completely satisfactory solution to remoteness does not exist, but one can achieve much by having a few selected items immediately available (desk-top library) together with a simple personal file and the use of available information services via telephone, telex and post.

7.3.1 Desk-top Libraries

At its simplest a desk-top library is a pocket library which consists of a diary or notebook. A few more items then can be carried on the desk top in the pocket, but the material must be just as easily manageable.

The criteria for inclusion of material in a desk-top library is frequency of use, availability of a local library and/or information service, and cost in time and money. As a guideline, if an item is used daily and is of negligible cost, it can be included. The items most frequently used by practitioners are a diary, local telephone directory, list of internal telephone numbers, dictionary (for report writing), and some items relevant to their particular jobs. The availability of the local library is the next criterion, and individuals need to determine their own patterns of use. A typical pattern is to visit the library once a week in order to carry out a planned series of tasks including scanning periodicals, verifying references, or looking up certain items. Detailed retrospective searches may be scheduled and carried out separately. Some information needs will develop which cannot be satisfied locally and yet cannot be left until the periodic visit to the library. People may ask their secretaries or assistants to find the information, or telephone colleagues or an information service, or visit the library; they may also need to make more provision for material within the desk-top libraries.

As for cost in terms of time, scanning from one end to the other should be sufficient to seek a known item. Selection should not take more than about thirty seconds which means a maximum of ten to twenty books and a small file of reprints or reports. Costs will also limit the size to this number of items.

Proper decisions on inclusion should be based on 'what is essential' and not 'what might be useful'. If the amount of material grows, one should honestly ask "When did I last use this?" Alternatively, two or more different shelves or files of material will be needed. Keeping material to hand for eventual need may be a poor substitute for a good card index.

In field work, ten to twenty books are too heavy to be carried around by an individual, but they can be carried on a shelf if there is sufficient accommodation. On site, it may be necessary to have more than ten to twenty books; if so, the material can be divided into more than one lot, each of which can be scanned in ten seconds or any time believed to be reasonable. The civil engineer on a construction site may keep the plans and associated texts together in one lot, perhaps in a set of drawers. Above them may be a set of reference books, including a standard handbook of civil engineering, the standard codes of practice, and possibly the trade catalogues of the principal products being used. A separate shelf may hold standard textbooks including items borrowed for a short period on a particular current interest, for example surveying. Beneath this, a second shelf may hold material on an objective-orientated topic of immediate concern, for example on welding structural components.

For students in their final year, some suggestions may be appropriate and are best made by subject tutors, or at least subject-specialist librarians. Postgraduates and practitioners will themselves have suggestions, and can become interested by finding obviously useful items which are not widely known. The information specialist can still provide guidance and, since users tend to hoard material, can suggest that collections are kept small and pruned ruthlessly. Another useful practice is to observe what material respected practitioners keep immediately to hand.

The bulk of material considered above has consisted of books and reference material for 'learning'. For people involved with specialized and current topics periodical articles and reports figure prominently. These tend to be more flimsy and less distinctive than books and can be more difficult to manage. Some sensible arrangement by topic may be possible, particularly for the research worker. Otherwise, it may be better to keep important documents in alphabetical order of author's name, together with a card index. For other items of potential interest a personal file may be appropriate.

7.3.2 Personal Files

The basic form of personal file is the card file.

Cards for files are available in a number of standard sizes, and bibliographical details of relevant material can be stored on them, particularly those of smaller size such as the cards often used in library catalogues. Most people will also record and store the essential features of the material, in which case the larger sizes will be better. The extent of these card files will naturally depend mostly on their potential value and usefulness. But users may also compile them, because the task of writing the essential details of relevant items as concisely as possible is good practice for technical writing.

Books and reports associated with cards should be returned to the main library after the essential detail has been abstracted. Reprints or copies of periodical articles should be filed either with documentation on a project or in box-files associated with the user's specialization. Additional material - eg. the records of the user's experiments, surveys, and other unpublished material - can be handled in the same manner, that is, by abstracting or recording location details on to cards and filing project material with project documentation and specialization material in designated files. Duplicate and fringe material, such as scripts prior to typing, can be disposed of fairly quickly unless the user is unusually methodical and has ample storage space.

One problem with subject files is the choice of headings. These headings may be more informal than those for group access in libraries, because they are intended for personal use and viewpoints. However, individual users do change their ideas, views, and search patterns, and the headings should be obvious to the user some time later; otherwise the preparation time will have been wasted. The recommended practice is for a few broad headings, each of which is either an area of specialization or a particular project. The headings may be imprecise and even disliked by some people, but they may be acceptable as 'catchphrases' and may have originated in the titles of significant conferences or other events. An alternative suitable for a specific list of headings is to use the headings in the index of a carefully chosen book.

One source of material can be the output from mechanized information retrieval systems, including selective dissemination of information services, which is often supplied in a card form. These cards may have headings, or numerical codes which suggest their own formal headings, and these are often detailed in printed indexes.

It is worth repeating the essential need for the files to be produced with the minimum effort and maximum benefits. Simplicity is likely to be an advantage; so is pruning, for example after completion of a project or publication of a report.

Some variations on the basic files can be considered. There may be a need for particular files, such as a set of cards which records the subject interests of a particular group. One part of the set is a list of subject headings, together with cross-references; the other part is filed in alphabetical order of individuals and provides details of their interests.

References can be recorded in book form either as part of a detailed diary, log-book or journal or as a book designed for reference. Detailed log-books are sometimes suggested for scientists and engineers but their value is part of the related work, eg. research. In comparison to card files, a book recording details has permanence and permits neat arrangement of notes or summaries of varied length. It is useful for certain specialists groups, notably students of literature. Its significant limitation is the difficulty of dividing details between files. Loose-leaf books serve as a compromise but have disadvantages. The usefulness of book records can be increased by careful indexing of the contents.

Records, including references, can be kept on edge-punched cards. These are similar to index cards but are pre-punched with a series of holes of about 2mm diameter set at a similar distance from the edge of the card and from each other. The holes on all cards in a particular set are in the same position. If that portion of a card between a hole and the edge of the card is removed, the card can be sorted from the rest by placing a knitting needle or similar object through the file of cards and shaking it. The 'clipped' card will drop out and can be perused. A system can be produced for relating each hole to a particular feature. Thus a file on people may have hole 'X' assigned to the sex of the individuals, clipping the edge for women but not for men. The card file can be easily sorted into men and women (and similarly into other divisions such as scientists and engineers). By using several holes it can be sorted into several groups, eg. physicists, chemists, biologists, metallurgists and engineers. The cards are useful for storing information, which can be retrieved by any of a number of facets, eg. the storing of analyses of steels by carbon content or any other designated element. Similarly if used by a large motor-vehicle stockist, a given model, colour or value, can be located if these features are designated. It should be remembered, however, that literature-search records do not necessarily need these multiple facets and that designing such systems, clipping the cards and sorting takes time and effort.

Currently some trials are being made of the use of computers to manipulate and store personal files. These are still beyond the means of most scientists and engineers but with increasing quantities of information and cost of manpower, and with improvements in computers the situation is likely to change.

Personal files are of particular interest to postgraduates starting major literature searches and to final-year undergraduates about to take up employment. Only the basic aspects need be taught to other undergraduates and to practitioners unless particular details are requested by the course tutors or students.

7.3.3 Use of Information Services

The essence of desk-top libraries and personal files is to keep them within the limits of information needs and linked to available information services. The more extensive and convenient are the information services, the easier it is for the user to minimize his file. However, the convenience of the services is limited by distance and telecommunications. It is often supposed that the ideal situation of a service is a library in the same building, easily visited even in bad weather; but such a resource should be viewed in the context of other key resources, especially time. A weekly routine visit to a library, with longer and more frequent use on appropriate occasions, should cause little problems if the library is available on the premises but not necessarily in the same building. There are, in fact advantages in having a centralized library service on the premises, not the least of which is that material for intensive use is ready to hand. Anyone carrying out an interdisciplinary literature search and using a fragmented library service will readily appreciate these advantages. Between the daily use of items in a desk library and a weekly visit to the library, there arise a number of occasions when brief items of information are required. Ideally the library's information service ought to be able to provide such items. The ideal is not always possible and there can be problems, not the least of which is the cost of providing such services. Another problem can arise when the responsibility for the accuracy and reliability of information is important. This implies adequate library staff in quantity and quality. However, to the individual, a telephone call to another information service several miles away is as easy as one to only several yards away. Provision of information services needs, therefore, to be considered on a regional or even a national level. Foreign and international information services can also provide assistance - preferably by telex or post, since they cost less than the telephone and create less likelihood of misunderstandings.

Thus, for each individual pursuing a worthwhile career and needing information, there is, or should be, a whole series of information resources.

1. Desk-top Library - intensive use (daily)
   - minimal in quantity
   - immediately available
   - relatively low cost and effort
   - supported by similarly limited personal files
   - provided and maintained by the individual for personal use within strict limits
   - contents comparable with such tools as slide rules, telephone, ruler, pen, micrometer, drawing board;

2. Organization's Library (if available)
   - regular visit (weekly) and occasional intensive use
   - contents accessible in minutes by visit or possibly less by telephone
   - optimal in quantity
   - expensive but key information resource
   - sometimes supplemented by active information service
   - provided and maintained by the library's organization
   - contents comparable with electron microscope, computers, and other equally expensive but usually desirable services;

3. Regional Information Service
   (Particularly important if the organization has no library)
   - sporadic use - often determined by importance of information need
   - size and scope often determined by demand
   - establishing a good communication link between service and clients is important
   - response sometimes slower than desired owing to limitations of operating such services
   - service may be mostly referral in that it indicates sources of information rather than providing it
   - often provided and maintained as a public service for the benefit of the community
   - service comparable with computer bureau;

4. Nationally Available Libraries and Information Services
   - occasional but sometimes intensive use
   - extensive and comprehensive collections but sometimes in limited subject fields
   - not immediately available, except occasionally
   - provided by government, professional institutes, corporate groups for their own benefit and for the benefit of bona fide enquirers;
5. International Libraries and Information Services
   - occasional use
   - collections and interest may be extensive (international) or specialized (linked to material of local interest such as a local crop)
   - response may be delayed by limitations of distance
   - services often provided for goodwill or for benefit of international relations.

7.4 DEVELOPMENTS IN INFORMATION RETRIEVAL

Some features of this Guide could be out-of-date less than a month after publication. Within one year of publication it is very probable that several, possibly minor, features will have changed.

Because of this many sections are 'philosophical' and do not give detailed instructions on the information to be provided or tools to be used. Although methods will need modification, it is hoped that the philosophy will remain reasonably constant. Attempts have been made to illustrate the philosophy in a manner that will assist the teacher. Thus details on many aspects have been added so the teacher can select those which best serve a particular situation.

In doing this it is important to emphasize methodology; the philosophy can be balanced by reference to specific bibliographical tools which can help them satisfy their immediate information needs. Our belief is that such specific instruction should be given where students are unable to satisfy their needs using the general methodology embodied in this Guide. The instruction can be given informally, as a result of students seeking advice from information specialists.

The closing session of a course on information retrieval can usefully be spent on the developments in libraries and information services. This enables the instructor to cover some of the unusual bibliographical tools such as 'peek-a-boo' systems and citation indexes, and to speculate on the future of information systems. It is possible to also include details of edge-punched cards at this point rather than under the heading 'Personal Files' (section 7.3.2).

One should accept that many aspects of librarianship are notably very traditional. In a world where arithmetic is done by mechanical means ranging from slide-rules to computers, where man has travelled to the moon, where conversation can be carried on between people thousand of miles apart, where a watch indicates the time accurately to within a small fraction of the part of the day, it seems that libraries are not very progressive. Changes are taking place, but major changes of direction are not easy. As an illustration, consider one type of bibliographical tools: abstracting journals. The usual search method is to take the most recent volume, look under a number of different headings and copy the details. This is repeated on the previous volume, and so on. One alternative is to have the details on cards, all the details that would have been scattered through many volumes can be grouped under any one heading. This slight advantage is offset by disadvantages such as increased cost and bulk and, from the user's standpoint, an unfamiliar presentation.

One development of the use of cards is the production of an 'inverted file'. Instead of a single card representing a single document it can represent a particular feature (topic or part topic). One particular form of inverted file which is convenient to use is known as the 'peek-a-boo' system.

7.4.1 Peek-a-Boo System (Optical coincidence systems)

Consider a file of cards with headings A, B, C, and so on. On the card headed 'A' there can be a list of numbers of relevant items in an abstracting journal or reference list. Thus the card can serve as an index. This can prove a particular advantage when the information sought has two or more facets, say 'C' and 'A'. In normal files (book indexes) one would look under both 'C' and 'A' unless certain of the indexing rules. With cards one simply seeks numbers common to both the 'term' cards representing 'A' and 'C'. There are devices for making comparisons easy. One of these is to print grids of several thousand numbered positions on each card, the appropriate numerical position being a punched hole for relevant numbered documents. Aligning the cards shows which holes are common to the two cards and the appropriate numbers can be noted.

To illustrate this one can take an example of an information need with more than one facet; the corrosion of aluminium by benzene. It provides three facets: aluminium (a), benzene (b) and corrosion (c). In theory, a comprehensive subject search would need examination of six different places in the index, as follows:

Heading	Sub-heading	Sub-sub-heading
aluminium	benzene	corrosion
aluminium	corrosion	benzene
benzene	aluminium	corrosion
benzene	corrosion	aluminium
corrosion	aluminium	benzene
corrosion	benzene	aluminium

Repeating this for a number of volumes becomes tedious and time-consuming. Matching three 'peek-a-boo' cards would indicate the total number of articles. If no articles were indicated, the search could be made less specific by removing one card e.g. the benzene card. If many more articles were indicated than needed, either the most recent number could be noted or a fourth facet could be added, say 'STRESS', 'HIGH TEMPERATURE' or 'REVIEW'. The result is a more complex system having two stages but with a gain in flexibility. Two such card systems (Geodex) are commercially available - for structural engineering and for soil mechanics. The basic cards with preprinted grids can be purchased from the publisher in the United States. If such cards are purchased for creating one's own files they should ideally be clearly printed with punching positions that can easily be matched and interpreted.

More sophisticated files can be created on magnetic tapes for computer searching with certain cost advantages, such as may result from the grouping of high-use terms.

7.4.2 Citation Indexes

When a scientist or technologist writes a book, periodical article, or report, it is common practice to refer to works published earlier. The main object of citing earlier work is to present a clear and fair outline of relevant work done previously so that one can assess the new contribution in context. Citations are particularly important in review articles and often unnecessary in manuals. As a result of this practice many scientists and engineers acquire a key paper relating to their own work and use the references given in the paper as sources of information. In using a key paper, standard text, or review, references more recent than that of the published item are needed as they can be very important. They can of course be sought in abstracting journals. This would be the typical procedure, but when a key paper is known there is an alternative type of bibliographical tool which is particularly useful where it is difficult to specify the appropriate subject terms for searching. This tool is the citation index. Two well-known examples are Science Citation Index and Social Science Citation Index produced by the Institute for Scientific Information in the United States. There are other well-established citation indexes in particular subject fields, notably law.

Citation indexes are produced by taking a very substantial number of published articles each year and processing them by computer so that in the index for a given year one can look up the 'key-paper' published earlier and note those papers which have referred to it in the current year. This processing is expensive and is reflected in the high price of the citation indexes. The Institute for Scientific Information has a co-ordinated group of activities which produces other useful bibliographical tools based on the same master files. This spreads the cost, but the other tools are also necessarily expensive.

7.4.3 Use of Mechanized Information-Retrieval Systems

The extent of course coverage of mechanized information retrieval systems must depend on the availability of such systems in particular areas. At one extreme, a few sentences will suffice but where mechanized systems are readily available and being used, substantial coverage is needed.

The following notes are intended for intermediate circumstances which may exist where mechanized systems are likely to become available in a year or so but details on access, subject coverage, availability and system commands have still to be decided.

Display charts or overhead projector transparencies can be used where material is available. If demonstrations or recorded audio-visual material are available, these should be used. Giving much detail in a handout is inadvisable when the presentation is more concerned with future possibilities than immediate use. But sometimes it can be helpful to issue publicity and descriptive material supplied by the data-base producers and their appointed agents.

A suitable starting point is a review of literature-searching as a manual operation. One variation of this approach is shown in Figure 7.4.3a, which lists methods of information.

Seeking Information from Literature

1st method - look through complete collection of material

2nd method - look through selected parts of a collection which has been previously organized
(USE CLASSIFICATION)

3rd method - look through lists of material, usually as an intermediate stage
(USE CATALOGUES AND OTHER FINDING TOOLS)

4th method - systematically search by perusing a selected choice of search tools, search strategy and search terms
(MANUAL INFORMATION RETRIEVAL)

5th method - specify information requirements according to formal procedure using selected files of machine readable information (data-bases), and developing search strategy and search profiles suitable to the system used
(MECHANIZED INFORMATION RETRIEVAL)

Figure 7.4.3a: Methods of Information

In outlining the development of machine-based information services, it may be useful to point out that the KWIC (Key Word In Context) index was the first use of computer-searchable files to develop an information-retrieval tool. This index was intended as a 'quick and dirty' fast alerting service for manual use. Very little intellectual work is involved in producing such an index and the user has to think of synonyms, related words, etc., when using it.

Machine-readable information files became available for searching in the early 1960's. During that decade the data bases were used in 'batch' mode to provide either current-awareness or retrospective searches. In the 1970's a different method of using machine-readable files was developed and is known as on-line access.

In batch-processing systems the organization making the search discusses with the enquirer his information need and translates this need into the vocabulary of the information store using subject headings, words that may occur in titles, etc., depending on what can be searched in the machine-readable data base. A search profile is then developed for the enquiry using logical connections of search terms that express the enquiry. All enquiries are then converted into magnetic tape and run together against the machine-readable data base. The resulting information or group of references is sent back to the enquirer when the search is completed.

The main use of batch searching is for regular SDI service where an enquirer regularly receives information as new material is added to the machine-readable data base. The user can subscribe to the SDI service and receive regular output just as he might subscribe to a regularly published printed index. It is also possible to use batch-processing to search a back file for retrospective information.

This type of batch search may be provided by the producer of the data base (i.e. the publisher of the printed index), or by special centres offering services in batch mode. Such services have been set up by national governments (e.g. in Canada), by nationally organized or regionally agreed service points (e.g. the Karolinska Institutet in Sweden which provides services in Scandinavia) or by universities or learned societies. Large industry and government departments may also acquire machine-readable data bases for internal searching.

In the 1970's it has become possible to search machine-readable files on-line, (i.e. for the user to interrogate a data base using a terminal or visual display unit connected by a telecommunication system to a machine-readable data base which can be many hundreds or thousands of miles away. With this method, the user may interact with the system as he receives its responses and can modify his search strategy as he develops his search. He can also search at one time the total amount of information on the file, which may amount to three to five years of store.

In the developed countries this method has led to the establishment of telecommunication networks that facilitate information transfer using the method. In developing countries, the nature of the telecommunication system may determine whether such a method of using machine-readable files is feasible or not. Where telecommunications are poor or almost non-existent, batch-processing may be the most effective way of using machine-readable information files.

Obviously the use of machine-based services is expensive and it is often difficult to justify the use of them. It should be realized that a search through a manual file is also costly in terms of search time and of providing and processing the manual file in the first place. However, these charges tend to be hidden and not identified in precise terms.

The introduction to the course session and any associated discussion should draw attention to the fact that many mechanized systems are still in the development stage. Equally important is the fact that, at present, mechanized information retrieval probably supplements rather than replaces other forms of information retrieval. Producers of machine-readable files and printed indexes do not plan to cease publication of the printed service at any time in the foreseeable future. Mechanized information retrieval has advantages and disadvantages and these are not always understood by new users. Some of these are shown in Figure 7.4.3b, a list which is simplified and subject to change.

Advantages

1. Offers an additional method of obtaining details of specific items of information from stored records.

2. Repetitive perusal of search headings and recording of possible items of interest is done mechanically.

3. Items distributed throughout the files having certain limited common elements (usually index terms) can be correlated and listed.

4. Searching time can be reduced.

5. Machine files may contain more up-to-date material than printed indexes.

   BATCH SYSTEMS

6. Production of current-awareness lists can be regular and routine.

   ON-LINE SYSTEMS

7. Search can be improved by interrogation of the system because of the interactive nature of the search.

Disadvantages

1. Payment of search costs may be unacceptable or a cause of difficulty.

2. Availability of systems may be limited.

3. Quality of output is highly sensitive to faults in construction of search profiles and associated work.

4. Familiarity with data bases and the systems appears to be significant in making the best use of information retrieval.

Figure 7.4.3b: Mechanized Information Retrieval Systems - Advantages and Disadvantages

7.4.4 Future Developments

Future developments are really little more than speculation, and these will change with fashion and innovations. Extrapolation of the rate of growth of literature makes one aware of future problems of processing and storage. Indications of the growth rate are sometimes illustrated by scaling up existing practices. For example, it has been said that by the year 2000 the complete population of the United States would be needed to produce Chemical Abstracts using today's methods.

The rate of growth of literature needs to be considered in the light of significant proposals for change in present-day procedures. Ideas sometimes put forward are for compact or more accurately 'compressed' storage using photography and, more recently, holography. Processing could be rapid by using automatic indexing matched with easy encoding of the needed information.

7.5 MODIFICATIONS AND CONSIDERATIONS ARISING FROM DIFFERENT CIRCUMSTANCES

7.5.1 Subject Differences

The basic structure is geared to undergraduate engineers. For different subject groups the following notes suggest some modifications and highlights.

General Science. In specialized information needs, data are likely to be very important and, although their coverage is already substantial, it can be further extended. In particular, attention can be given to two aspects: data associated with unusual phenomena, and comparative evaluation of available sources of data handbooks. When there is an unusual phenomena, there is an increased possibility of:

1. the information not being available;
2. the information being less reliable because evaluation and comparison are less likely;
3. there is a need to consult other sources in addition to data handbooks.

If a single feature is unusual, the data (phenomenon, material, condition) are likely to be available in published literature, but the searcher may need to make his own evaluation of their accuracy and reliability. If two features are unusual, published figures is much less likely but still normally worth some search effort. The likelihood of evaluation is very low, but assessment may be possible by comparison with associated data. For example, if the thermal conductivity (usual feature) is needed of gold containing 20 per cent nickel (unusual material) at $300^{\circ}C$ (unusual condition) then evaluation may be based on the series of determinations of the thermal conductivities at different temperatures by the individual worker. Examining a graph of the values, and considering the apparatus and techniques used, give the guide to approximate confidence limits for the scientist.

As to sources of data other than data handbooks, certain abstracting journals can be used, authoritative bodies can be consulted, and finally, if the potential value of the information justifies it, the value can be determined by experimental methods. Most of the larger abstracting journals which cover chemistry or physics provide data comprehensively within the abstracts.

The other important need is to be aware of associated work on similar topics, both prior to starting research and during the work itself. Registers of research are the ideal method, but their range and coverage leave much to be desired and suffer from publishing delays and infrequent revision. Contact with people working on associated topics and use of current-awareness techniques are necessary to minimize the risk of unnecessary duplication of research, which can be very frustrating.

Mathematics and Computer Science. Data and statistics can be important in practical applications. Current awareness may be especially important in the relatively fast developing field of computer science. Current developments in libraries could be of interest because of attempts to quantify and qualify various library interests and operations and also because of the use of computers to manipulate large amounts of information in libraries.

Astronomy. The astronomer is likely to have particular specialist information needs, which are based on specialized data and factual information. Particular reference works are likely to be important and it may be appropriate to include in a course material suggested by a subject tutor or specialist. This particular area of information may also involve current awareness based upon scanning selected named journals (again, guidance from subject tutors or specialists can be useful). These particular needs will be reflected in an interest in personal files and desk-top or observatory libraries.

Physics. Data are likely to warrant much attention and current awareness may be important. As with science in general, knowledge of work done elsewhere and personal files will be of interest.

Chemistry. Chemistry has a large and well-structured literature with notable bibliographical tools. Discussion of bibliographical sources may be more worth while than emphasizing specialized information needs. Standard reference works, such as Beilstein's Handbuch der Organischen Chemie are comprehensive and need to be understood for full use. Indeed, the same applies to Chemical Abstracts, which includes an excellent patent concordance. If course discussion is based on sources rather than needs it may be useful to bring in a subject tutor or specialist and stimulate an informal dialogue.

The structure of the literature and the nature of the field are likely to create interest in current-awareness techniques and personal files.

Geology. Two special characteristics of geology are the particular types of specialized information needs and the possible remoteness of libraries during field-work.

The specialized information needs will be based on maps, and a sample collection can be considered a library resource.

In the field, planning will be important so that information needs are mostly anticipated. Services should, if possible, be specific, eg. manuals, and it may be worth creating a simple personal manual if a satisfactory published one does not exist. A closely analogous problem is the handling of samples, which can be weighty and bulky, and which need transporting to a central laboratory. Whatever provision is made for conveying samples may be modified to include communication of information or literature.

Personal files and personal literature collections are likely to be of interest.

Life Sciences. Botany and its related subjects are similar to geology, possibly involving a remote area of work and specialized information needs. In other respects biology may be comparable with chemistry.

Certain types of factual information, including both data and statistics, may be of interest.

Engineering and Technology in General. Section 7.1 on specialized information needs is orientated towards engineers.

Engineering data need adequate coverage. Patents often have a particular interest but are not given the same coverage by all general bibliographical tools. The information contained in patents has particular value and is not easily acquired from other sources. Acquiring information from patents is a specialized task because of the nature of the material and the complex indexes and classification scheme.

Materials Science and Technology. As for chemical engineering, except that certain specialized information needs require more detail. The obvious category is materials information itself. In some cases, notably steels, a substantial number of variables can affect values of mechanical properties. Thus a precise value for a particular sample can only be determined by tests. The literature can often supply typical properties, and perusal of these for a given material will show the wide range of variation. If the material is the subject of standards, then acceptable limits will be published; but it is important to realize that standards normally give limits and not typical properties. The limits can often be easily achieved. Conversely, one must not assume that materials even to standard specification always give typical values.

Additional specialized information needs exist, such as equilibrium diagrams and isothermal diagrams for steels. Appropriate specialized bibliographical tools exist for these as well as coverage in one more general bibliographical tool.

Mechanical Engineering. Product information can be relatively important. Problems may exist because name and subject indexing may be inadequate for certain areas of interest, such as mechanisms. Materials information is significant and elaboration is useful; it could include comparative costs particularly related to strength or weight values.

Civil Engineering. A number of particular features are worth noting: specialized information needs; the remoteness of many sites from central services; and the seriousness of some failures.

Of the specialized information needs, attention should be given at least to data, statistics, product information, standards (including codes of practice) and materials information. The remoteness of sites can be offset to some extent by telephone and postal systems. If the company's own vehicles travel between sites and central services, use of these can be made.

In the construction industry, at least one company has found it beneficial to use its information services to 'vet' all tenders and plans and so reduce some failures. This has been a result of the information service being frequently used after an incident, the common remark being "If you had asked us, we could have told you how to avoid it". Thus identification of information needs has been improved by 'vetting'.

Aeronautical Engineering. Some of the specialized information needs and the current-awareness techniques are important. Coverage of report literature and provision of information services tends to be relatively well-developed.

Automotive Engineering. As for mechanical engineering and engineering and technology in general.

Production Engineering. The specialized information needs are fairly obvious: data, statistics, product information, standards, materials information. A particular problem area is the need for cost, materials and process information in which relationships of cost/strength and cost/weight are itemized. Indeed, one has the impression that the literature lacks the mundane but significant category of information on alternative processes and materials, so that choice of material, manufacturing process, and level of finish are determined more by experience and intuition than by use of literature.

Electrical Engineering. Certain types of data are important, as are product information and standards.

Remoteness from centralized services is sometimes a problem for the practising electrical engineer, but he is probably in a minority.

Electronic Engineering. Specialized information needs include circuit diagrams, for which some reference books are available. Otherwise at least one abstracting journal (Electrical and Electronic Abstracts) indexes appropriate terms.

Current-awareness techniques are important and need full coverage. The field is relatively fast developing and the period of maximum use of periodicals is relatively short.

Chemical Engineering. Specialized information needs should include data, statistics, product information, standards, materials information, commercial information, and authoritative organizations and individuals.

Current awareness techniques need coverage, as does the working situation. The chemical engineer is not normally remote from library services.

Medicine. Specialized information needs, current awareness, personal files and information centres are important. Time can be a significant factor.

A suitable approach can be based on established and understood practices and extended into other information needs and techniques.

Agriculture, Horticulture, Forestry, and Land Management. People in these fields are probably working in remote locations and possibly without centralized services. This may be balanced to some extent by low frequency of significant new developments and slow progress resulting in a low number of information needs beyond the core knowledge gained through education and experience.

Food Science and Technology. Contains elements of chemistry, agriculture, biochemistry and chemical engineering and must draw on the schemes suggested for these subjects.

Of the specialized information needs, the important ones could be factual information, product information, standards, news items, and authoritative organizations and individuals. If the library facilities are limited, authorities on particular topics can be the most important (e.g. FAO).

For each of several specific crops (such as coffee), there appears to be a centre of excellence and expertise which can provide information

There is a vital need to communicate information very quickly in cases of fires, diseases, and pests.

Management. A notable feature of management is the very wide range of specialized information requirements, which include all those listed in the basic outline of this Guide, plus management-information systems. The latter are developments for recording and controlling the flow of information which is generated by the organization for its own use. Mostly it relates to processes, finance, sales and purchasing. The subject also tends to be both multidisciplinary and interdisciplinary and not surprisingly the workers in this field are often one of the most frequent users of information services.

Also important are commercial information, current-awareness techniques, desk-top libraries, personal files, and available information services and developments in libraries.

In short, the basic course structure is well suited to management with the possible addition of management information systems. Such systems have strong communication links and communication and information-retrieval lectures could be satisfactorily linked by this common topic.

The literature for management studies is not well structured and is possibly best approached through study of information needs.

Social Sciences, Humanities and Arts. The basic course of instruction starts with general principles and philosophy, followed by the basic methodology, and in the final session detailed studies. The bulk of the final session is unsuited to social sciences, humanities and arts except where features such as news items are covered. It should be noted that Section 7.6.2 deals with official government publications.

Higher Education (Undergraduate level). The basic **course** structure is aimed primarily at scientists and technologists within this group.

Postgraduate Education. Postgraduates, like social scientists, may have needs so different from the basic structure that it is advisable to start by specifying the educational objectives of their course. If the teacher is new to the teaching of information retrieval and is at a loss how to produce a basic outline for this third session, a seminar can be held using the basic structure as a check-list.

The reasons for difficulties with postgraduates are the level of their specialization and depth of the study. Some people believe that an undergraduate on completing his studies should be more skilled in obtaining information in his own speciality than the general information officer or librarian.

The depth of knowledge in the specialization has usually been gained at some cost in terms of general knowledge, and some science departments believe it desirable to include non-scientific questions in the exercises for postgraduates.

Much more thought and research is needed on the teaching of postgraduates and techniques of particular interest can include case-studies, games and simulation. The first technique increases collective knowledge and can be realistic. The same is true of the latter two techniques, which can be useful in correlating various key resources - time, information, manpower, material and finance.

In-Service and Further Education. Specialized information needs can be described according to the subject background of the group. Current awareness is important for practising scientists and engineers, but the emphasis should be on available resources. Similarly, information 'at the desk/in the field/on the site' is important and instruction should include practical solutions to typical problems. Developments in libraries will normally be of interest only for bringing students up-to-date on what is available.

The emphasis should be on available techniques and resources and in practical terms with the minimum of philosophy. Games and simulation can be as useful here as for postgraduates.

For short courses, serious consideration should be given to reducing this third session to a bare minimum and using an extended practical session in which specialized needs are covered by the practical exercises. Individual attention can be given to the more receptive students who readily comprehend the basic techniques. For this purpose numbers attending short courses should be limited, ideally to about twenty. Experienced staff can handle up to thirty students on these sessions when they last for a maximum of two hours.

The basic outline assumes numbers greater than this and is appropriately a compromise.

7.5.2 Differences in Teaching Approach

Formal Approach. The basic structure indicates a formal approach in the form of a lecture. This is suitable for large groups, particularly those with a common subject interest. The objective of this final session is to extend the basic methodology taught earlier. Teachers having knowledge and experience of the specialized areas are desirable and a suitable approach is the use of clear handout notes, supported in the lecture by interesting illustrations. Feedback and questions should be sought so that clarification can be given. Audience participation can be very useful, as some of the more convincing illustrations come from contemporary experiences. The teacher can use 'planted' members of the audience to stimulate others, but this may be felt to be too informal or even misleading. There is a risk of tedium in the third session, because a number of specific aspects are quickly considered and described. Some stimulus to overcome this is very desirable, but difficult to achieve in a formal approach.

Informal Approach. In a relatively small group an informal approach, even a discussion or seminar, can be used, so that the basic structure becomes a check-list for discussion. Handout notes can be used to counter the risk that students retain insufficient detail to be of real use. With postgraduates and practising scientists and technologists, valuable contributions can be made by the students themselves.

Practical Approach. The specialized information needs, current-awareness techniques, and vocational information systems are probably best taught by practical methods in the ideal situation. Unfortunately, ample time is not usually available. The nearest possible solution is to include an appropriate range of questions in the practical exercises, supplemented by handout notes. In groups of less than twenty the subject material can be incorporated during supervised practical sessions. However, the basic structure should be used as a check-list in a post-practical discussion.

Theoretical Approach. The theoretical approach is negligible and is covered to some extent in the basic principles and the information-science approach.

Information-Science Approach. It is possible to consider various information needs and the extent of individual searches as particular variations. Thus a current-awareness search is normally made over a period ranging from one week to three months, but a retrospective search usually covers a number of years. The search statement for current awareness is usually generous and extensive compared with the specific and precise requirements for retrospective searches.

There is a tendency to view systems in terms of the extremes of reference retrieval and data retrieval. Reference retrieval does not supply any raw or primary information (which is the real need of scientists and engineers) but only secondary information i.e. where the primary information is located. Data retrieval normally leads to physical, chemical, or engineering data and statistics as the output. Thus a scientist asking for a particular fact is provided with precisely that. There is a whole series of intermediate situations. The distinction between data and statistics can be difficult, for example whether the rate of flow of a river should be regarded as a datum or a statistic. Thus, the output from a particular retrieval system can be secondary, primary or intermediate information.

Possible Forms of Information Provision

- abbreviated reference
- full reference (including title of article)
- reference plus added descriptors
- reference plus descriptive abstract
- reference plus informative abstract
- reference plus extended summary
- reference plus full article
- extracted/interpreted information
- comparative/correlated information
- evaluated information

Librarianship Approach. The librarianship approach, i.e. describing the available literature, can be applied to the specialized area of interest described in the basic outline as specialized information needs.

The librarianship approach can be seen in the suggested structure for the development of search techniques. It is most apparent in the sections on current awareness, where the available techniques are considered, in desk-top libraries and personal files, and in developments in libraries. It can be extended by reference to specific available services for current awareness. It can also be extended, by slight modification, to specialized information needs by considering, for example, trade publications rather than product information. In most cases, this is not recommended (exceptions are official publications and certain subject fields, notably in social sciences, humanities and arts).

More Extensive Approach. It is possible to extend, by including more specialized aspects e.g. official publications; by having information specialists run subject tutorials throughout a student's period at university; by ensuring that specialized skills are at an appropriate level; and by providing materials or resources which encourage self-development.

Some desirable results may be achieved by providing assistance (e.g. a current-awareness service) to postgraduates and final-year undergraduates so that on leaving the university they recognize a loss of service. Eventually training may be given on computerized systems during an appropriate part of university studies (postgraduate and final-year undergraduate projects).

More Intensive Instruction. If the instruction needs to be intensive, such as a one-day short course, a printing of the basic outline is suggested together with greater emphasis on practical sessions for groups of up to twenty students.

7.5.3 Differences Arising from Locale

The basic outline for the third lecture on development of search techniques will need modification according to local circumstances, especially in developing countries. However, it is difficult to generalize about these circumstances.

In both education and practice in developing countries the attitudes are likely to be very 'down to earth' because the range of objectives is practical. General and specialized information-retrieval lessons should reflect this attitude in both context and presentation, including the questions posed in practical exercises.

In discussing current-awareness needs, it may be appropriate to describe the three alternatives in reverse order, since scanning needs the minimum of processing or effort by libraries and selective dissemination of information requires computers and their associated software, which may not be available in a particular country. It is possible to produce useful current-awareness bulletins if ample manpower is available. It may be worth encouraging coverage of the country's own recently published material. The extent to which SDI is described should be related to its availability and plans for its provision.

When considering 'at the desk/in the field/or on the site' information systems, a teacher may secure a great deal of attention, especially from practising scientists and engineers, who are alert to possible solutions of their problems. A simple solution does not exist, but one can achieve much by good personal preparation and planning, by having telecommunication and postal links and by acquiring or even creating a manual. This aspect of instruction may be worth expanding at the expense of other aspects in order to cover particular subject groups. If the group is mixed, a useful illustration can be the information needs of a small boat at sea. Good preparation is vital to safe completing of the voyage, and anticipation of information needs is part of good preparative. The need for information can be urgent and important; passing an unfamiliar buoy on the correct side; identifying a navigational aid; understanding a signal from a large vessel. Some form of accessible and easily digested check sheets in waterproof coverings are often suggested and constitute a form of manual. Telecommunications can be useful to small boats.

7.6 OTHER IMPORTANT TOPICS

7.6.1 Report Literature

The contents of some reports can be very significant and can have wide implications; they may be published later in periodicals. However, it has been realized that in developing subject-fields access to reports can circumvent delays in publishing and can provide information of immediate importance. In some highly industrialized countries this has led to the use of reports as a large-scale means of circulating information, covered both by specific bibliographical tools and by existing abstracting journals, etc.

For the information user wishing to search report literature, the first choice, if it exists, is published guides to report literature. Alternatively, he can locate and list report-publishing bodies in the relevant field by consulting appropriate reference books and examining, if possible, lists of reports from these bodies. Or he can contact or visit the authoritative organization in the subject-field and consult any of its unpublished indexes or browse through its collection. The latter is likely to mean additional work for the authoritative body outside its normal terms of reference. Thus such use of facilities should be kept within reasonable limits.

7.6.2 Official Government Publications.

Governments have a commitment to inform people of the limits (laws) which exist for effective development of the community. Furthermore, they may believe they have a commitment to inform the people they represent, of their actions and opinions. This can be done through publications.

Scientists and technologists may need to consult this literature for various items of information such as:

- enquiries on incidents of scientific and technological interest;
- maps, charts, handbooks, reports from government authorities in particular fields;
- legal information;
- codes of practice e.g. on temporarily closing a public highway.

They are best served if there is, readily available, a published guide giving details of the publications. If there is no such guide, a handout can be produced for course purposes. A lecture can incorporate brief details and illustrations, which can include a simple flowchart as given in Figure 7.6.2.

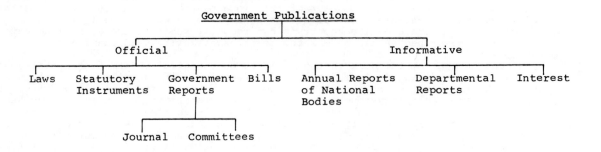

Figure 7.6.2: Forms of Government Publications

7.7 CHECKLIST USING THE APPROACH GIVEN IN THE GUIDE

1. Have you established the types of information needs and literature which are important to your specific groups and which require coverage additional to that in the basic search techniques?

   These may include: Data
                         Statistics
                         Product information
                         Standards information
                         Patents information
                         Legal and medical information
                         News
                         Commercial information
                         Theses and dissertations

2. Have you adequately covered current-awareness aspects of information retrieval?

3. Have you covered the difficulties, frustrations and opportunities of obtaining information in remote areas?

4. Have you given sufficient attention to expected changes in information retrieval which can affect individual information-seeking habits? Do you know what time-scale you are using as a basis for instruction? Do you have a satisfactory balance between immediate and long-term coverage?

5. Have you identified the range of viewpoints, needs and literature for the common subject interests of your group?

8. Practical exercises

Practical work is believed to play a key role in developing basic **information-retrieval** skills - second only to the role of determining satisfactory educational objectives in the total scheme of instruction. Because the setting of practical work is fairly easy, there is a real risk of giving it insufficient time and attention and, if assistance is available to the teacher, it can be usefully employed here. However, those assisting should be given clear and precise instructions on the nature of the exercises.

It is arguable whether a teacher should start with the practical exercises and build the lecture sessions around them, or whether he should base the exercises on the instruction. In reality, the exercises need ample preparation and both they and the lecture material should be developed together. The exercises should be based on the real needs of the group and should not require detailed specialist knowledge. Both, lectures and exercises have a common starting point: the information needs and the available literature.

The basic scheme outlined in earlier chapters incorporates thinking about practical work. It is useful to review the three basic approaches:

1. Simple exercises based on principal information needs.

2. A correlated series of exercises based on a realistic topic and requiring a range of approaches.

3. Stepped exercises, ranging from simple to more difficult.

There are other approaches, but they can usually be related to these three.

8.1 SIMPLE EXERCISES BASED ON PRINCIPAL INFORMATION NEEDS

This general-purpose approach has been successfully used with many groups, including undergraduate students. The basic technique is to determine as far as possible typical and common information needs for the group. For each type of need at least one question is set, and it is usually made as simple as possible as well as being based on significant finding tools. For key types of information need - which may be referred to as introductory texts, articles on specific topics and data - a second question can be set - more specialized and difficult, and slightly more realistic. Information needs requiring advanced specialized skills should not be included because they would be found to be more difficult than other more important types of need. Figure 8.1 shows typical examples, designed for use in the United Kingdom. The position of questions may be found illogical, eg. the use of 'articles on specific topics' occurs only in questions 6 and 7. A more logical position is following the introductory publication question (Q 1) as it is likely that one would expect books to be suitable for that and periodical articles for the others. However, it lies where it does because students find the use of abstracting journals difficult and time-consuming, and are more confident and capable after having successfully completed the first part of the sheet. In short, one tries to embody some of the advantages of the other approaches to practical exercises.

This type of practical session, if used properly, can develop a reasonable range of skills which are satisfactory for both educational studies and project work, and can do so with the minimum expenditure of effort and time. It has disadvantages and it is important to overcome them as far as possible. Since the exercises must be capable of completion within a reasonable time it is not possible to introduce more than one question on each basic type of information need. This suggests adequate briefing or guidance and also avoidance of over-simplified questions such as the name of a local taxi firm. It also means a special effort to develop skills to the optimum level. This can be achieved if the students' departments set appropriate projects and the information retrieval teachers are available for subsequent guidance.

PRACTICAL LITERATURE SEARCHING

NAME ....   COURSE ...GENERAL ENG............................

QUESTION	ANSWER
1. Trace an introductory text on "Photoelasticity in engineering materials"	AUTHOR: TITLE: PUBLISHER: DATE OF PUBLICATION: SOURCE OF INFORMATION (& page):
2. Provide the bibliographical details of a book entitled: "Energy for Man: Windmills to Nuclear Power"	AUTHOR: PUBLISHER: DATE OF PUBLICATION: SOURCE OF INFORMATION (& page):
3. What is the density of Aluminium?	 SOURCE OF INFORMATION (& page):
4. What is the recommended safe working load on a single vertical polypropylene rope with a diameter of 18 mm?	 SOURCE OF INFORMATION (& page):
5. Trace an organization concerned with metal sinterings (in U.K.)	 SOURCE OF INFORMATION (& page):
6. Give details of a recent article on government proposals for industrial safety	TITLE OF ARTICLE: AUTHOR: PERIODICAL TITLE: VOL.NO: ISSUE NO: DATE: PAGES: SOURCE OF INFORMATION (& page):
7. Give details of a recent article on structural changes in quartz glass during heat treatment	TITLE OF ARTICLE: AUTHOR: PERIODICAL ARTICLE: VOL.NO: ISSUE NO: DATE: PAGES: SOURCE OF INFORMATION (& page):
8. Give the name and address of a manufacturer of aluminium plate	 SOURCE OF INFORMATION (& page):
9. Give the number and date of the British Standard Specification for industrial safety belts and harnesses	 SOURCE OF INFORMATION (& page):
10. Trace figures for the U.K. consumption of zinc sheet and strip in 1971 and 1972	 SOURCE OF INFORMATION (& page):

Figure 8.1 : **Example of Practical Question Sheet**

The production of the exercise sheets can introduce problems. A standard duplicated blank sheet can be produced specifying the details of the information required, such as publisher and source of information. The actual topics can be taken from significant finding tools so that each question is answerable. The questions and answers can be noted in card files. If all students had the same question sheets, many would work collectively and, unless subsequent discussions were enforced, this would defeat the objectives of the exercises. It is possible to arrange the sheets so that students who have the same question 1, for example, do not have another question which is the same. Typing the topics on the standard sheets is tedious and time-consuming and, if typing assistance is limited, this can produce problems.

The question sheets should be taken seriously by the teaching and subject departments. They should be marked and returned to the student via the departments. If designated as course work, the standard of completion is usually very high. This is generally an advantage, but can create problems by increasing the demand on certain finding tools. A watch should be kept in order to ensure equality of opportunity to consult these tools.

8.2 CORRELATED SERIES OF TOPIC-BASED EXERCISES

This approach has been used mainly with postgraduate students and undergraduates already having basic information-retrieval skills. At these educational levels the exercises tend to call for a greater degree of specialization and for topic-based work, together with fulfilment of professional roles. The exercises are therefore intended to reflect the specialized interests and the realistic needs that are then becoming apparent.

Figure 8.2 shows a typical example of a correlated series of exercises. The starting point is the selection of a range of topics. Earlier departmental projects and topical subjects provide ideas. For each topic a range of realistic information needs are thought out and specified as questions. The examples given can be completed in one and a half to two hours, which is as much time as most students can afford to spend on exercises. The questions are ideally distributed for a supervised practical session so that performance can be related to available time and guidance provided as and when appropriate. In many cases individuals can work collectively and between them provide a substantial amount of information on the various topics.

The disadvantages of this approach to practical exercises are:

1. The lack of overall coverage of the main information needs.
2. The varying degrees of difficulty relating to the choice of topic.
3. The need for prior exercises or a large amount of guidance appropriate to the degree of difficulty.
4. The general difficulty of assessing performance with such a range of variables.

Earlier use of a set of simple exercises can overcome these disadvantages by giving a common basic skill, which can then be assessed, which can provide for realistic and useful experience and which can be developed using this topic approach. But attempts to combine a simple range of exercises with realistic common topics does not often work. It is an abnormal topic which itself embodies all the basic information needs and it is extremely unlikely that one can construct a series of topics with a common level of difficulty.

A good source of topics for university students is the projects used in the subject departments and in comparable universities. The departments' information on these may be a little 'dated' and an awareness of new developments in the subject fields would be useful.

With postgraduates it may be thought desirable to let the individual students choose their own subjects and then to develop a series of exercises on the chosen themes. The resulting exercises can motivate the students but it may prove difficult to produce question sheets in the available time. Some questions can be quickly dismissed because the answers are already within the students' sphere of knowledge.

Given experience, it is possible to produce a number of realistic questions associated with a certain topic. If this is done without working backwards from known publications, the result is usually realistic. A particular question may not be answerable, and this adds to the realism. However, one should attempt to ensure that at least some questions are answerable, in order to prevent discouragement. Some course tutors believe that at least one question should be unanswerable and that another should be based on information outside science and technology.

The varying degrees of difficulty and the inclusion of questions not based on topics may cause concern. A simple solution to the first of these problems is the inclusion of a particularly time-consuming and skill-stretching question as the last question on the exercise sheet. For the second problem, the non-topic-based question can be related to the students' non-topic needs, based on professional activities or marital, trade-union or other outside interests.

With certain groups it may be desirable to add supplementary questions as the practical session proceeds. This is useful in correcting judgments, for example if the existing skills have been underestimated and for providing supplementary information, either on request or if existing skills have been overestimated. Taken further it can introduce simulation of actual working situations and influence the practical exercises as the session proceeds. This is comparable to an instructor feeding information into an aircraft simulator during the development of flying skills. The simulation approach is very attractive but can be time-consuming and difficult to prepare. More work and trials are needed to show whether it is a viable approach.

Experience has indicated that basic skills should exist or should be developed prior to topic-based exercises. These may have been imparted through a simple range of exercises as suggested earlier, or by using a variation - eg. a modified question. The use of a variation has been found of value, particularly if the basic exercises have been used some time earlier. In developing the basic skill it is necessary to develop a simple approach in which the answer is obtained. In variations this can be recalled and then extended by stressing the location of material by more than one approach. The advantages of such modified questions are better use of time, emphasis on a different choice of finding tools and approaches, and interaction between tutor and students. The disadvantage is that the motivation of hunting for an item of information and assessing its reliability and relevance is lost. It is therefore appropriate to include in a variation some questions which seek information rather than source lists. A question can be included - such as working plans of a particular steam locomotive or a photograph of an iceberg - which is not easily answered using traditional search techniques.

8.3 GRADED EXERCISES

Learning itself is a skill and is to some extent based on practice. Short-course students may be out of practice and inclined to practical work rather than lectures. In addition, practical work may be more convincing than any soundly argued case. For such groups the use of graded series of exercises may be appropriate. The initial ones should be relatively simple, to build up confidence, and the subsequent ones should develop and extend existing skills. Although they require a lot of time, such exercises are perhaps the real essence of any serious development beyond the basic information-retrieval or information-management skills.

As a start it is useful to consider a one-day course for in-service scientists and engineers, with group sizes restricted to limits imposed by the number of available finding tools and information-retrieval tutors. The group members are likely to be in unfamiliar surroundings and to need to orientate themselves to the subject, their associates, and the location.

A brief introduction to the objectives and background can lead to a short visit to the library, and the first series of exercises can be based on the more familiar finding tools such as the principal encyclopedias, standard dictionaries, telephone directories and library catalogue. The second series of questions can be a series based on information needs in a broad subject area, and so on.

The principle of graded and repetitive questions can be extended to many other situations. Students may even ask for additional question sheets. Development beyond the basic course given in the earlier chapters can be based on further practical exercises, with supporting description and discussion relating to the exercises. One could, for example, alternate between practical exercises based on common topics and a series of exercises initially extensive, and specializing on particular areas such as data. Two hours appears to be a suitable time to complete a practical worksheet and description, briefing and discussion can be carried out easily in one hour. With highly motivated groups supervised sessions organized informally allow individuals to become really involved in principles and techniques.

For these extended courses some useful guidelines can be found in exercises used in the training of librarians for reference and information work. But it is necessary to differentiate between training librarians in exploiting their collection and providing a service, and training users to find information efficiently. The difference is not great, but it is important when the users are scientists and technologists.

8.4 OTHER APPROACHES TO PRACTICAL EXERCISES

With experience it is possible to try other approaches, many of which are variations of those already described. It is to be hoped that newer approaches will lead to more effective teaching.

Simulation is attractive but demands resources, time and effort, which may not be justified by the benefits. It is interesting that library and information schools do not appear to use simulation to any extent for the introduction of different information needs.

At the other extreme, it is possible to produce hypothetical solutions to exercises without using material. There is little advantage in doing this other than as an examination question which emphasizes information retrieval and provides some measure of students' knowledge.

Figures 8.4a, 8.4b and 8.4c show variations on exercises which have been used.

ENGINEERING DESIGN CENTRE   LITERATURE SEARCHING

PROJECT 721   Design of a large lawn mower for greens and golf courses.

Manufacturer's specification demands:

i. That the mower be hydraulically powered and self propelled, but as light as possible so that cutting can be accomplished with minimal damage to turf and green.

ii. That the mower cuts with a traditional cylinder action to a width of 20" or more.

1. The starting point is the possible adaptation of a gang mower. Is there a book available which would provide a brief description of a gang mower - with an illustration if possible?

2. Trace a directory listing manufacturers of gang mowers, so that one may be purchased as the basis for a prototype model.

3. Provide a periodical article describing the application of hydraulic power to large lawn mowers.

4. Trace a specification for agricultural mower parts which could be incorporated into the design of the new model.

5. Are there any fairly recent figures available on the value of the lawn-mower production industry in the U.K.?

6. Large rotary mowers are a possible source of competition. Is there a market survey on these, listing models available with prices, advantages etc.?

Figure 8.2 : Example of Practical Question Sheet

Practical Exercises

Name ....   Group ....

PRACTICAL LITERATURE SEARCHING

1. Give the author, title, publisher and date of publication of a book which describes the properties and uses of hardwoods.

2. What is the density of African mahogany?

3. Give the name and address of a local saw-mill.

4. "Hermattan sometimes splits tree-trunks". What is Hermattan?

5. Give the name of an exporter and importer of hardwoods.

6. What is the principal source of balsa wood?

7. Give the name of a wood which is heavier than water.

8. What is the name of an international organization interested in forestry?

Figure 8.4a : Example of Practical Question Sheet

Name ....   Group ....

PRACTICAL LITERATURE SEARCHING - REVISION

In the following questions indicate three or more sources of information which you regard as relevant and advisable in the particular cases. As an example, question 1 has been completed.

1. The telephone number of any local glazier

    Sources of information: 1. Trade section (Yellow Pages) of the telephone directory
                               2. Local newspaper - Agba Times
                               3. Ring public library on 786-1122

2. Textbooks on irrigation

3. Articles and reports on high-yield wind-resistant cereals

4. Annual rainfall, including variations, in the western provinces

5. Cause of Dutch elm disease

6. The name of organizations funding international research and supplying information on soil fertilization

7. Suppliers of ploughs rope-hauled by stationary engines suitable on dense impacted clay soils

Figure 8.4b : Example of Practical Question Sheet

RETURN TO LEARNING   LITERATURE SEARCHING

TOPIC .... You are invited to choose a topic
of interest to you and to satisfy as many of the following as
possible:

1.   Details of an introductory publication or book

2.   Details of a recent document or periodical article giving current
      views or facts

3.   Recent data or statistics relevant to the topic

4.   Names of **organizations** with relevant published views or
      manufacturing relevant products

5.   A British Standard or other official publication providing the
      authoritative view or standards

Figure 8.4c : Example of Practical Question Sheet

9. Additional considerations

For the most part this Guide is based on an ideal situation, which rarely exists. In practice the teacher may be short of time, subject to pressures of other work, and ill-adjusted to the needs of a particular audience. The students may be tired from substantial lecture commitments, solely concerned with obtaining a good degree, more concerned with immediate matters, their careers or recreation, or from backgrounds in which literature has played a negligible role. The classrooms may be ill-equipped, uncomfortable and badly designed. The library and its resources are subject to economic and other problems, and this may cause gaps in the available finding tools. Whatever the difficulties the teacher must aim at a methodical and logical approach to a collection of material, even if it cannot be treated methodically all of the time.

The basic outline must be modified according to variations in local situations. Minor modifications to the approaches used in this Guide are given as notes at the end of the relevant chapters. Sometimes more extensive changes are necessary - discarding prepared lecture notes and giving an extemporary presentation.

Cases needing drastic changes can be dealt with by intelligent anticipation and are most likely to occur in a typical situation. One example is the visiting lecturer, particularly one visiting a new country. Developing countries often have rapidly changing circumstances, and hence need particular attention.

It should also be stressed that situations demanding alternative approaches do not necessarily arise because of any lack of quality in either teacher or students. Greater attention to information-seeking in secondary education can make the more general contents of this Guide largely redundant for subsequent teachers. However, lack of enthusiasm among students is likely to be normal; information retrieval and library use are not associated with fields of study in most students' minds. The real problems occur with mixed groups and with a cumulation of unusual factors requiring particular attention.

The general solution to such difficulties is to use practical approaches, smaller groups, more preparation, and involvement of both the group and individuals.

9.1 STUDENT DIFFERENCES

9.1.1 Group Size

In general, small and large groups have already been catered for in earlier chapters. Extremes may necessitate an alternative approach.

One extreme is the case of a single student or groups of two or three. Beyond this number the general outline can be used.

With one to three students it would be ludicrous to present a lecture of any type, least of all the formal type. One should have a two-way discussion plus practical work. The students should talk as much as, if not more than, the teacher. Obviously the actual events will be determined nearly as much by the students as by the teacher. A start may be made as follows:

> "We need to start by discussing the value of information.
> Have you any experiences where something did not go right
> because you did not have the information you needed?"

The session can close with practical work based on a topic chosen by the students or meeting an immediate need. The second session can be a follow-up in the library. Finally, it may be of value to set an essay on the subject, thus ensuring that the student finishes up with detailed notes and providing feedback to the teacher, who, of course, can comment usefully on the essay.

The more formal lecture can be used with groups of up to 120 students - if necessary, up to 180. A change in approach is required where the group can only be accommodated in a large lecture theatre equipped with full-sized projection screens. Here individuals beyond the first two or three rows cannot be identified, and the session becomes less of a lecture or lesson but more of an educational programme.

For the teacher this approach is likely to involve the use of other people in supporting roles and careful attention to rehearsing or developing his own role. The degree of support and rehearsal will probably depend on the actual size of the audience.

A projectionist is useful, even for groups of a hundred or fewer students. Slides are best prepared by professionals or talented amateurs. If well-prepared, they usefully illustrate certain points and also arouse and maintain interest. It is helpful if the relationship between engineering projects and the use or non-use of available information can be demonstrated. Photographic evidence of success or failure is not usually too difficult to obtain.

A chairman or other person may give a personal introduction, and it is often useful for him to establish contact with the audience, using, if necessary, the excuse of seeking information - for example on audience numbers.

With audiences of several hundred or more consideration should be given to a range of supporting roles, some of which have their equivalents in the entertainment profession. For example, apparatus can be wheeled on and off by stage technicians. One possible case is the use of games which contain a relatively high element of entertainment. Thus, members of the audience can compete against each other, either singly or in teams, and with or without access to reference tools. The teaching element lies in showing that information can be of value and that there is actual skill in getting the right information at the right time. Such techniques are not for inexperienced lecturers, however.

The educational content can be increased by colleagues fulfilling particular roles or presenting specialized aspects of the subject.

Perhaps the largest audience is the one which watches a filmed or televised presentation. For the inexperienced presenter this is likely to be a recorded performance, and weak points will have been improved during production. The differences between a large audience and the camera are not so great. The camera is impersonal and does not provide substantial feedback; but it does allow re-runs and pauses. Problems arise when trying to handle a live audience and a television camera together. Advice is sometimes given that tutors should concentrate on one and ignore the other.

9.1.2 Educational Level

This Guide has referred mainly to university and college students. Some comment has been made about teaching information-retrieval skills prior to university; this should not be too difficult for people used to dealing with these groups. The material already given may serve as a basis, suitably modified or simplified for the students or pupils and the resources available to them. Thus the child who has learned to read is likely to find it useful to locate material on his interests. The older pupil doing simple experiments should be able to relate the results of his experiments to those published and determine some causes of the difference; this is a fundamental part of the scientific approach. Of course, if basic information-retrieval skills are developed prior to the students' arrival at university, a problem of determining a suitable lesson content arises for the university teachers. A similar but not identical problem already exists for the person faced with developing information-retrieval skills among university teachers themselves.

The wise approach is to determine the extent of knowledge by exercises and questions and to develop useful skills related to gaps in existing knowledge. The result can be a mixture of exercises and discussion, with active participation by all members.

For students who already have the advantage of earlier instruction, it may be beneficial to develop use of the fairly extensive collections of universities. For university teachers the emphasis can be on the more recent scientific developments. Those who are confident of their existing knowledge may reconsider it when faced with exercises and the testing of their own hypotheses.

Examples of specialized areas about which many library users are ignorant and would be grateful for direction are: personal indexes, citation indexes, patents and statistics.

9.1.3 Unresponsive Students

The most common problem is likely to be unresponsive students. A wise teacher will expect to meet the situation. On a new course he is likely to meet it because the basic material has not yet been subjected to continued reassessment. The inexperienced teacher may find response fades in and out, according to circumstances. Poor response is an indication that something is amiss. The main possible causes - teacher, lesson content, audience and environment - should be examined immediately, in case there is a danger of attendance falling.

There are two important guidelines here - to avoid the problem by anticipation and to avoid mistaken interpretation of student responses. The latter is a very easy mistake to make. Concentrated attention can give the impression of being the opposite; an experienced listener may **relax**, sometimes slumping on a desk or leaning back with closed eyes. He is deliberately concentrating on the key aspects of the presentation and cutting out distraction and detail.

The first step is to consider oneself - the teacher - one's tiredness, the amount of preparation, one's understanding of student background. If tired, one should try to use the tried and **tested** approaches with which one is familiar. One should concentrate on essentials, seeking by questions to ensure that each point is understood. If the replies are satisfactory, one can proceed. It may be that the lesson finishes earlier than intended, but an early finish is not in itself a bad thing.

If the replies are unsatisfactory, either more than one element is out of tune or one is extremely tired. In the former case one should consider other possible reasons for lack of response. If the audience is large and a tape/slide presentation or film is to be shown, it can be viewed earlier in the session and while it is on the teacher can examine his approach. Alternatively, if difficulties have been anticipated and a colleague is prepared and present, the teachers can be switched. If one is simply over-tired, one should try to stay with the main features of the prepared material and give the best possible summing up at the end. This should include a statement of the objectives of the lesson and should give an opportunity for questions.

If the lack of response is thought to be due to the teacher's lack of preparation the solution again may be to stay with the familiar material, but to seek oral questions so that one gains feedback from the audience. If the cause of the problem is that one has run out of prepared lesson material, one should finish the lesson unless a very good reason exists for continuing it. It is better to lose 20 per cent of lesson time than 50 per cent of the students for subsequent lessons. When a series of lessons is involved, it is important to give priority immediately after one lesson to preparing for the next. If lessons for two groups are consecutive, they both need preparing in ample time. In an extreme case, where each day is full of lessons, the day should finish only when the following day's material is ready. But ideally, unless one is an experienced teacher with several years' training, one should endeavour to limit oneself to three hours' lecturing a day. An obvious exception to this rule is a short course, when one may have a group of students for a day or more. Preparing such a short course may take more time, and it can be an advantage to involve more than one teacher. If one curtails a lesson, the next session may naturally follow - and lack of preparedness is avoided. However, the proper supervision of practical exercises is itself a demanding task involving teaching. Tiredness and lack of preparation together are unsatisfactory.

Other long-term aspects of preparation may help with short-term preparation problems. In some parts of the world there is current interest in newer methods of teaching and scientific work. In teaching, apparently simple cases are used to develop a scientific approach.

A typical case is to produce a number of simple structures within stated limitations using rudimentary materials. These are then tested to failure and there may be a small prize for the winning team or individual. An alternative is to produce a method of crossing numerous obstacles of the same type, say ditches or canals. Various teams have produced simple devices for harnessing solar energy and wind power. A wide range of groups will react to these basic problems, and knowledge that there is published information on such topics is valuable in creating the appropriate response.

The problem is finding suitable cases; but it is one of information retrieval, and there is a reasonable choice.

The third factor concerning the teacher is the lack of awareness of the students' background, especially of differences in subject interests and in culture. Subject background is a matter of preparation, including discussions with course tutors. Ideally, the novice teacher, or any teacher in unfamiliar circumstances, should start with groups which share with him a common subject.

The cultural problem is a little different. Whereas attitudes and backgrounds may differ for cultural reasons, they are linked by the skills of finding information by using libraries. The solution to cultural differences is to minimize the background and theoretical studies and to use practical exercises. These exercises should be drawn up after consultation with course tutors or other people familiar with the local situation. The lecture sessions can become question-and-answer sessions, together with briefing for practical exercises. For the visiting teacher a few days for familiarisation can be very helpful. So can, talking to early arrivals in the classroom.

So far we have examined one possible factor in poor response - the teacher. The other factors are subject content, the audience and the environment. For the subject content one needs to check new and previously untried material - the occasions when it proved suitable; and the relationship between the content and the needs of the audience.

As to the audience, it is the difference between this group and those at other sessions which is important. One course of action is to ask the audience what they are finding as the obstacle to attentiveness, but this can seriously worsen the situation if not done properly. It may well be that the answer is a matter of acoustics or comprehension. Another approach is to have a question-and-answer session based on an appropriate aspect of the subject. For example, "how valuable do you think the resources of a library can be to an engineer?" (One needs to be ready for the answer: "Not at all", and be prepared to follow with another question.) Such an approach is best with a small group, particularly of mature students. With large groups of bored students there is a temptation to give some entertainment - to combine personal involvement with checking the reason for poor attention.

The environment obviously influences the audience. Distracting elements, such as temperature or lighting, can be easily spotted.

9.1.4 Unconvinced Students

Unconvinced students can be a different problem from those who are unresponsive, but the two groups are often associated. The teacher can be out of tune with the needs and background of the students, in which case lack of conviction follows lack of concentration.

Students may be unconvinced for a variety of reasons. Scientists and technologists tend to be objective in their thinking; and this can result in a very narrow specialized viewpoint. Fortunately some educationalists now favour developing thought processes in different ways, so that individuals are more flexible in their thinking and can adapt themselves more easily to different circumstances.

None the less, one can expect at least some members of a group of scientifically-orientated students to have fixed ideas and be very single-minded. A teacher of information retrieval is unlikely to be fully accepted as teaching a valuable aspect of science or technology. This applies not only to students seeking good qualifications, but to more established scientists and technologists, particularly those with practical responsibilities.

These attitudes can intensify if teachers are not apparently connected with the scientific profession but follow a different vocation. The title of the course can also be discouraging, a classic example being 'A Lecture on the Use of the Library by a Member of the Library Staff'. 'Information Retrieval', 'Acquisition of Information' or 'Finding Out' are often more appropriate titles than 'Use of the Library'. Departmental staff may introduce the course. To be introduced as "...an engineer who has now worked in libraries for a number of years and who will teach you some skills by which you can find information for forthcoming projects, in connection with various other aspects of the course, and, indeed, for your eventual jobs" is a very good start. Part of the problem is of course that not many engineers or scientists have worked in libraries for many years.

It is hoped that the suggested course structure and material are sufficient for most scientists and technologists. In a number of cases this will not be so. Those retaining their lack of conviction will fall into two groups: those who are unresponsive and have not properly considered the subject and those who have considered the details but do not accept information-retrieval skills as significant. In some circumstances the interests of the majority of students will prevent attention to the few who lack conviction. It is therefore useful to arrange subsequent discussions (or even tutorials) with them, singly or collectively.

The solution for both those lacking interest and those who do not accept the significance of information retrieval appears to be participation and practical illustrations. For those lacking interest, especially the younger ones, interesting scientific cases involving information retrieval can be used. These may be major scientific and technological developments, such as space technology and deep sea studies. When appropriate the illustration can be related to problems of immediate concern or recollection. A student who has struggled with and 'fudged' for a design, essay, piece of equipment or experiment, and is suddenly presented with information that can give a satisfactory solution, is easily convinced in a matter of moments. Close co-operation with the students' departments helps enormously in the choice of illustrations and the teacher's own subject knowledge, if appropriate, can help substantially. There is a danger in presenting ideal solutions to students' problems in that some may believe that information will normally fit their needs precisely in this way. Students have on occasions complained that an otherwise perfect article, report or book is not suitable because the units of measurement are not the ones they prefer or need to use. Similarly, complaints have been made when two items of literature have proved necessary to cover all the relevant aspects. Fortunately such niggling protestations are infrequent.

9.1.5 The Special Case of Production and Maintenance Engineers

Supplying information of immediate use or relating to an easily recalled need is a very satisfactory method of convincing the older experienced scientist or technologist. It is likely that he will be involved in a certain amount of repetitive work, often with substantial responsibility and often concerned with numerous short-term problems. People concerned with production and maintenance in industry often fall into this category. The classic example is the maintenance engineer, who is often required when things are not going well and often develops a blunt and out-spoken manner, which is usually linked with high standards of workmanship and expertise. Such characteristics arise from the nature of his job; on the one hand overcoming oversights in design, and on the other hand dealing with people who complain about equipment or persuading them that a certain product, idea or technique will make life easier for them. The competent members of this group often rise from lower levels, sometimes assisted by so-called engineers who are not suitable for 'other' engineering duties such as design. The experienced and competent maintenance engineer can be a good source of inspiration for the teacher of information retrieval. If a teacher has a group containing engineers of this type, he must structure the session to their practical needs.

The central information services usually attempt to show their value to this category of engineer by illustrative cases. Often this is done by an exhibition, static or mobile, in which broken farm implements, new feed stock, improved machine tools, etc., are used to show how available information has improved the performance, benefits, efficiency or reliability of a particular exhibit. This approach can be usefully adopted by the teacher of information-retrieval skills. The engineers can participate in discussions about how improvements to machinery can be made, and the role of literature can be shown. In practice many of the engineers will indicate that their source of information is often other people - "it's not what you know, it's who you know". But it can be shown that somewhere along the line of communication someone often consults the literature, which is thus exploited indirectly. It is useful to demonstrate the direct relationships that exist between every-day problems and solutions in the literature. This will at least gain the attention of the audience; it will not encourage them to visit libraries because they have to work where the problems exist; but they may be prepared to add librarians and information officers to their network of contacts. Another result of such instruction is that certain items of literature may appear in workshops, production offices, and so on. It is a useful practice, when visiting factories and workshops, to note what books and documents are in use.

It is not necessary to bring components into the classroom when engineers need to know detail. Working drawings are sufficient, but in many cases reference to typical situations is all that is needed. It is essential to emphasize the link between problems and solutions in the literature as stated above and the practical steps that a practising engineer can take. An extreme example is a ship's captain, one person who is unable to leave his job in order to find a piece of information in a library. However, the wise captain carries the appropriate charts, tide-tables and current-awareness notes on recent sunken obstacles, buoyage systems, etc.

A teacher has achieved something if he can show that intensive use of a few sources may lead to the use of other material. If as a result the work of the engineer is benefited by one or two additional reference books and assistance from capable information sources, the teacher has reason to be satisfied. Convencing a group of engineers of this type is not easy and is a task for the more experienced teacher, particularly one with experience in what engineers call 'trouble-shooting'.

For the teacher without detailed background in this area, it is advisable to do some background reading and then talk to relevant people and organizations. Unfortunately the literature is scarce, because those with pertinent experience rarely contribute to it. The 'publish or perish' attitude does not exist in this area, and there is little prestige in 'how-to-do-it' papers. Writers of motor-vehicle workshop manuals may make a living from their work, but their publications will not impress most interview boards.

Some case histories are published, and these are useful. The first-year student may be impressed by major scientific achievements. But the practical engineer is more impressed by mundane difficulties: the cooker control panel which melted when turkeys were cooked for a festival; the fountain pen which leaked because ink corroded the breather tube; the tiles which fell off the swimming pool because the adhesive was wrong; or the shop-fitting crew who cut the job time by 40% when they borrowed an explosive bolt gun of a type which had been on the market for two years.

The other area of reading is on common topics such as 'speeds and feeds' for machine tools, particularly those normally controlled by machinists. Engineers associated with these common and widely used tools will appreciate accurate and reliable data. There are several sources of 'speeds and feeds' information. The common source is the machinist, who judges the circumstances against earlier experience. In a few cases co-operative bodies have collated information on 'speeds and feeds' from different sources and produced a data-bank. However, such data-banks are not likely to have been established in many developing countries.

Further sources of information are the manufacturers of machine tools and the cutting tools, and these may be very important in the less literature-conscious industries. Some surveys have shown that the able technical sales representative can play a key role in the transmission of information, though one must differentiate between the knowledgeable and able person and the glib but less able one. Manufacturers can also supply useful information for other categories of engineers such as design engineers. Product information is one example. Here the situation is well developed by some design engineers, but individual manufacturers and suppliers tend to produce what is known as 'trade literature'. This can range from glossy brochures indicating the superiority of the product to detailed handbooks. Teachers visiting developing countries, or indeed any other country, are advised to determine the basic nature of trade literature in that country. Other areas of common interest are preventive-maintenance schedules and general-service procedures.

The needs of production-orientated engineers have been dealt with at length. They are important but often overlooked. They are increasingly significant in developing countries, particularly those seeking self-sufficiency in products which have hitherto been imported. Domestic consumer products are an example of this. It is worth repeating the basic instruction technique which appears to work with this group, i.e. numerous case studies based on real examples. Repetition will reveal patterns of approach acceptable to individual engineers. At this stage some engineers may ask questions about the reasons for using particular approaches. This is a sign that some of the basic theory can be given - preferably in answers to specific questions.

9.2 SUBJECT DIFFERENCES

The differences between the arts and humanities, social sciences, and science and technology may be based on false distinctions but they have become established and each has its own logic and methodology. In the ideal situation a trained engineer specializing in the literature and information will teach engineers, and so on. This ideal cannot always be attained. An organization may need only one engineering information specialist, who may be faced with teaching medical-information retrieval. Even more likely, other information specialists may be faced with teaching engineering-information retrieval.

This problem has been discussed elsewhere in this Guide and detailed ideas have been presented. The general solution is more effort and time spent on preparation, especially in studying the background and needs of the students; a greater emphasis on practical aspects and case studies; and greater participation by the students. This usually means smaller groups for particular sessions. Sometimes joint sessions with subject teachers can be useful.

9.3 NON-SCIENTIFIC SUBJECTS

Different approaches are needed for non-scientific subjects, and several possibilities have been referred to very briefly in this Guide. Two subjects stand out particularly - librarianship and education. In the former, information retrieval exists as a subject in its own right and needs to be expanded much further. Our experience has shown that it is useful to include library students in user instruction, since they appreciate its practical orientation. In the case of education students it is believed that a similar, though not identical, case exists. Teachers in pre-university education ought to consider the implications of teaching the acquisition of information. Maybe one should teach the instruction of information-retrieval skills. At least, a study of the potential benefits should be carried out.

The following notes may serve as a starting point for teaching students in librarianship and education.

9.3.1 Librarianship

Three basic points need to be taken into consideration:

1. Librarians and students of librarianship have their own information needs, which can be satisfied by exploiting the literature. In this they are no different from students in any other vocation except that the literature on the subject has its own characteristics.

2. Abnormalities and special cases exist in any significant group of information, for which routine approaches do not work. The usual course for library users is to consult librarians. Librarians can, of course, consult other librarians, but they need to go deeper in special cases. Naturally, the subject tutors have their own material and guidelines, but experience indicates that practising and teaching members of the profession have different views, which complement each other.

3. One can argue that information retrieval is part of engineering or any other subject. This is useful for motivation, but one should accept that motivation can exist for students of librarianship and, indeed, librarians. It may be of a different type, because individuals may believe either that they know it all or that their subject tutors will provide them with the necessary guidance. If motivation appears to be a problem it is worth setting exercises or examples which fully extend the students' capabilities. The differences between teachers of librarianship and practising librarians should be recognized.

The approach adopted can be information retrieval, information management, library use, or a combination. Perhaps the last approach is the most useful. Its content can be based on surveys of library use, and the occurrence and identification of information needs, together with alternative sources of information. The remaining background material is hardly needed and the same is probably true for material on search techniques. Exercises and material on special cases of particular information needs or categories of literature can be expanded.

9.3.2 Education

It is difficult to give specific recommendations about instruction on the teaching of information acquisition in a wide range of educational establishments. One can teach information retrieval to educationalists in terms of their study and vocational needs and in terms of provision in schools.

In teaching teachers, their best interests are probably served by the library use approach. That is, use of the bookstock of the library, the bookstock available through the library, periodical literature, etc. This does not mean that detailed lists of useful sources need to be compiled. Taking recreational management as an example (associated with physical education), recent reference books are available which substantially change the approach that might have been used a few years ago.

9.4 NEW SITUATIONS

The purpose of this Guide is to improve the quality of teaching information-retrieval skills to users of scientific and technical information. Those teachers with experience in this role will already have learned by trial and error many of the key criteria for success. This Guide should be of maximum value in situations that are new, particularly for the teacher. Special consideration needs to be given to the more common of these new situations:

1. New to teaching information retrieval to users;
2. new types of groups for instruction;
3. new locale for teacher, groups or both.

Each of these situations can be considered separately, but some teachers may find the situations in combination. If so, they should consult all the remaining parts of this section. They should also find it advantageous to work in co-operation with one or more colleagues, plus local liaison and assistance.

9.4.1 New to Teaching Information Retrieval to Users

The best advice is to gain experience by working alongside an established teacher of information-retrieval skills. One should also make ample preparation, use tried and tested techniques, and start with predictable groups.

Ample preparation should concentrate on the following priorities, in order:

1. Establishing educational objectives;
2. determining nature and characteristics of group;
3. setting goals;
4. producing draft programme;
5. producing practical exercises;
6. producing lesson content; and
7. scripting first few minutes of the lesson.

The Guide has advice and comment on these priorities, often in detail which can be difficult to digest unless ample time is given to it. Often it is better to complete one stage in preparation and continue on other work before proceeding to the next stage.

Tried and tested approaches are based on techniques for locating publications (books), recent articles (documents) and data. The current emphasis is on discussion of reasons for developing skills and on the logic of the techniques. A common method of describing finding tools is by handouts, but superior methods should be developed to ensure self-sufficiency without resorting to sheafs of papers. Several enthusiasts strongly support the use of case-studies. Having ensured proper content and approach, the teacher has the important problem of developing and maintaining interest. Case-studies and participation are good ways of tackling this, but the former is often specific and the latter can be time-consuming. Handouts can reduce these disadvantages to some extent.

Students are probably the most predictable groups, the individuals having a similar background of age, subject of study, and interest. Opportunities usually exist for teachers to sit in on other lessons and obtain background details. Often the locale will already be familiar. Once lectures have been established for a course, they will remain to some extent similar from year to year.

Difficult groups are those whose make-up, beliefs and background cannot be anticipated.

9.4.2 New Types of Groups for Instruction

The needs and background of some groups can be successfully predicted: student groups on established courses are a good example, and for this reason the content and main presentation has been directed towards them. Groups of practitioners and other outsiders are more difficult because they can have varied reasons for attending a course and it is difficult to predict their responses and attitudes before the course.

For short courses of practitioners and others the content of instruction is adequately covered in a logical manner in Chapters 5 to 8. The difficulty arises in selection of the appropriate content and the method of presentation. Selection of content will largely be determined by the educational objectives, which in turn are determined by the needs and background of the members of the group. Thus, selection is partly a matter of finding out about potential group members and partly a matter of experience. Finding out about group members can be difficult, and it is necessary to peruse publicity literature and work closely with course organizers.

Presentation can be difficult and, if other circumstances add to the difficulties it is advisable to keep group sizes small and concentrate on practical exercises and discussion. Project-based approaches can be particularly suitable provided that the topics of the projects are relevant to the interests of group members. Preliminary exercises offset the other disadvantages of project-based approaches and exercises. Project-based approaches are most easily used with postgraduates and older undergraduates.

On a one-week course for mature people returning to learning a successful approach has been based on a short lecture plus an hour's practical preparation and the option of two to eight hours' practical literature searching. A typical group size is fourteen, which gives the ideal ratio of seven users to one teacher when two teachers are used. The option of from two to eight hours permits special study of a skill of the individual's choice, and the majority tend to choose literature-searching. The key stage of the course is practical preparation. The individual course members can be given the option of a mixed selection of basic questions (Figure 9.4.2), basic questions on the topics of their choice (as previously shown in Figure 8.4c), or extended literature-searching practice based on a project agreed between member and teacher. Those wanting a basic set of questions can continue literature-searching with guidance, if they wish. Extended literature-searching techniques based on topics result in sheets like that shown in Figure 8.2. The result tends to be complex but effective, for once the individuals appreciate the options, they can choose a pattern of questions suitable for themselves. Because of the limited time and the complexity, it is advisable to gain experience before attempting this individual approach for unpredictable groups.

LIBRARY STUDIES POSTGRADUATES   LITERATURE SEARCHING

Name:_____

Please attempt to answer the following questions which might well be received by an information service:

Notes:

A) To add to realism, one question is virtually unanswerable.

B) Information officers in the university library may be consulted in the role of colleagues and enquirers.

C) Give details of published sources consulted.

D) Do not consult external organizations but, where appropriate, please indicate the names of relevant organizations who could be consulted in the event of failing to find the information.

1.   How many strings are there on a zither?
      Can a sketch or plan be provided of the instrument?

2.   What is the specific heat of potatoes?

3.   Where can Turner's painting 'The Fighting Temeraire' be seen?

4.   How many students are there in teacher-training colleges in the U.K.?

5.   Is there anything published on water-divining for farmers?

6.   Provide, if possible, the name and address of a service which will search for water on farms.

Figure 9.4.2 : Example of Practical Question Sheet

Appendix 1

Introductory tours of the library

One of the most common forms of instruction in library use is the introductory tour of
the library for university and college students. A tour may range from a five-minute
stop on a tour of the campus - with the guide saying "Books to the left, journals to
the right" - to detailed preparation by the library and teaching staff using aids such
as tape/slides, films and video recordings. The important points are: avoiding confusion
between information-retrieval and library-use skills; appreciating the individuality of
a particular library in, for example, loan procedure; and linking the structure of
introductory tours to an objective purpose.

1.1 OBJECTIVES

The possible objectives include:

(a) indicating the location of the library building;

(b) indicating the internal arrangement of collections and services;

(c) indicating where material on a particular subject is located;

(d) publicising the library;

(e) creating initial student/library-staff contact;

(f) adopting the practices of other libraries;

(g) increasing real or apparent use of the library;

(h) informing potential users of the procedures of borrowing, locating and using
material;

(i) interpreting the symbols, words, numbers, signs and codes on catalogue cards
and elsewhere;

(j) drawing attention to good bibliographical tools;

(k) creating interest in the library as a source of information;

(l) examining material relating to various aspects of writing, printing, bookselling,
binding and general librarianship;

(m) correcting false impressions about libraries;

(n) enlivening the library staff;

(o) taking the opportunity to develop one's interests in people;

(p) developing the ability of the staff to enjoy talking to students;

(q) supplementing other instruction and guidance;

(r) having an excuse to avoid other more boring or difficult work.

The contents of this list, which are not complete or mutually exclusive, show the wide
range of possible reasons for having introductory tours of libraries. It is useful to
ponder on some of the items listed. Some may be seen to develop the right technique for
wrong reasons, some may seem to be very virtuous. The consequence is that introductions
are likely to vary considerably, particularly when one considers the differing sizes
and arrangements of different libraries.

It is often suggested, rightly, that if an introductory tour is given, one must
accept that it will not achieve the ideal in present circumstances. Even so, one should
endeavour to reduce it to essentials which cannot be better supplied by alternative
means. In other words, its primary objective is to make up for deficiencies in students'
knowledge.

When considering objectives the following factors should be taken into account:

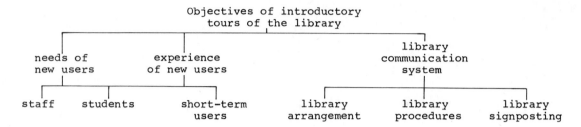

Figure A.1.1 : Library Tour Components

1.1.1 <u>Needs of Users</u>

The needs of a user will depend on the potential and likely use he or she will make of the library and his previous experience of library use.

<u>Academic Libraries</u> - The two main groups of users are the staff and students of the university or college. New members of staff may need to provide their address and collect borrowers' tickets, and so it is useful for the library's information-service or reader-service staff to invite them to visit the library and be shown features of interest.

By comparison, the intake of students is large, and an individual approach impractical. Collectively, new students can find university or college a new experience and may be bewildered by the relatively large size and range of activities. The library may find itself in apparent competition with the student union, the central administration, departmental administrations, course tutors, personal tutors, health services, residential services and catering services.

Timing is important (discussed in Section 3.1.2), and relates to need. At some point the students will be expected to submit their first work which may need the use of the library. The introductory tour should precede this event.

Postgraduate students should be guided as well as opportunity allows. It may be possible to treat research students in the same manner as staff, and advanced-course students may receive a good deal of guidance associated with supervised exercises in information retrieval.

The needs of individuals can be met simply by the provision of an enquiry desk and an evident and consistent readiness of senior library staff to assist.

Two real problems exist with first-year students. One is the time available for meeting different needs, and the other is uncertainty about the degree of guidance needed. Students want to use the library for many different educational purposes, which arise at different times. There is, for example, a contrast between the need to borrow books which have been placed on reading lists and the need to borrow reports on inter-library loan. As to the degree of guidance needed, there is the experience of a university library which set pre-university students simple exercises using a range of material. To their surprise there was very little consultation of the staff standing by to give guidance. This experience contrasts with the plentifulness of routine enquiries on the location of material in a library. A practical solution to this problem appears to be the provision of adequate permanent guidance together with an introduction concentrated on a few essentials such as readiness to help. For permanent guidance some libraries have printed publications giving a guided tour. These are worth trying, but it takes time and effort to make them attractive and yet avoid too much detail or flippancy. If humour is desired, it can sometimes be achieved in illustrations to a serious text.

Academic libraries also have "short-term" users, usually people on short courses. Introductory tours are not needed for these unless they are attending a short course on information retrieval for practising scientists and technologists. On a one-day course twenty minutes can be generous enough for a tour of a small library. The bulk of the day will be spent studying finding tools. For larger libraries or longer courses an hour is more appropriate. These are general indications, only, for circumstances can vary - for example, when the library has a particular feature such as an on-line information-retrieval system. In general it is wise to allow a few minutes for orientation and then decide on the amount of additional time according to reactions. People wanting to start their exercises should be allowed to, and those wanting more library detail can be provided with an extended session.

<u>National Libraries</u> - These are usually large with a range of users. A well-written guide and an efficient enquiry desk are likely to be appreciated and may be sufficient.

<u>Public Libraries</u> - The smaller public libraries probably have self-evident arrangements for enquirers and assistance. The larger public libraries have enquiry desks and may organise library tours for particular groups.

<u>Special Libraries</u> - The librarians and information officers of special libraries should be receptive to users and be prepared to assist new members of staff and explain the arrangements for stock and services.

A suggestion has been made that teachers of information retrieval in academic libraries should endeavour to provide some continuity after students leave university. This can be done by maintaining close links with special libraries and information centres and providing a mechanism for introducing new potential users and services.

1.1.2 <u>Prior Instruction in Information Retrieval</u>

With notable exceptions, potential library users are unlikely to have had prior instruction.

Practising scientists and technologists will have experience of problems associated with their subject fields and, once convinced of the value of published information, are receptive to library instruction. It is useful to have some teaching on location of material in the library before embarking on practical exercises.

Some scientists, technologists and others, particularly those with teaching experience, may have made considerable use of libraries, which means that they are self-taught in library use. For them a visit to the library for competent library users is simply a method of orientation and assessment of the stock and services.

Postgraduate students may have received instruction in information retrieval and library use as undergraduates. The information-retrieval course will thus be part revision, part up-dating and part extensive instruction.

Among new undergraduate students earlier experience and instruction in library use is likely to vary considerably. Some will have made substantial use of libraries and may be over-confident of their ability. If so, they should make library tours in small groups so that individual backgrounds can be taken into account. Small groups also allow greater flexibility in approach and personal contributions by members of the groups.

Teachers on a formal course of instruction will place importance on evaluating any prior instruction. This is necessary if the course itself is to be evaluated.

1.1.3 <u>Library arrangement, procedures and signposting</u>

These aspects may be called collectively the 'library communication system' - the means of providing information on the use of the library and receiving the views of users in return.

There are two aspects of the system. One is the procedure for making use of the library's stock and services, and the other is the location of specific items within the library.

The simplest method of indicating the procedures is to describe them in an easily digested form which is readily available. The obvious method is a printed library guide or handbook, which is distributed to each reader and up-dated often enough. Additional copies should be available within the library. If a guide is to be used, it should be concise, well produced, and properly indexed, together with a contents page. It should include library regulations, a map of the library, opening times, the division of the library stock, the terms of borrowing and copying, and a brief description of the library services available. A description of certain features, notably the library catalogue, may be appropriate, and certainly the library classification should be named, if not described. A printed tour can be included or kept separate, according to preference.

Locating specific items in a library may not be easy, particularly if the library signposting is inadequate. To become more familiar with the arrangements it is worth taking any opportunity to visit libraries and judge their arrangements. For the purpose of this Guide one can concentrate briefly on the library catalogue and indexes, stock arrangement and library signposting.

The library catalogue and indexes can be dealt with in the library guide or handout and possibly in the form of notices as well.

Stock arrangement is a matter of logic. The common practices are worth adopting and some standardization of library practices would be useful. Books are divided between loan and reference sections, and within the former there can be separate control of books in substantial demand. Other subsections are rare books and books too flimsy or too large for the shelves. One can add to these the various treatments of periodicals and a further collection for specialized forms or content of material, e.g. official publications or statistics. There is a tendency for some users to overlook certain subsidiary collections, such as important conference proceedings and some major annual reviews. The library can provide details of its stock arrangements in the library guide or handbook. It can also have the library services and bibliographical tools placed next to the entrance.

Library signposting is not practised to the extent that it might be, probably because of a desire for an attractive layout rather than a clutter of notices reminiscent of a multiple store. (In fact, the objectives of a library and a multiple store are similar.) However, signposting within a library ought to be carried out with the care appropriate to a place of study as well as a store of literature and information.

1.1.4 Operational Objectives

These are, in effect, a list of the intentions of the introductory tour, which can be measured, if appropriate.

Considering a group of undergraduates studying science or engineering and having been part of the university for a few days, the following operational objectives could be appropriate:

(a) Awareness of the existence of the library;

(b) knowledge of the location of the library;

(c) knowledge of, or ability to find, the proper location in the library of any item listed in the catalogue, periodical holdings list, or similar stock listings;

(d) knowledge of, or ability to find out about, possible courses of action if the required item -

  i) is not stocked by the library;

  ii) is stocked by the library but not on open shelf;

  iii) is stocked by the library, but a loan copy is required;

(e) General impression of useful reading material and readiness of library staff to help.

Awareness of the existence of the library would seem to be easily achieved. But a visit to the library is likely to give substance to this awareness - as well as a knowledge of its size, range and location. The problem of creating awareness of departmental libraries must be solved according to circumstances, including the needs of readers.

The third operational objective - knowledge of, or ability to find out, how to obtain a required item - is more complex. The library guide can embody many details, including a flowchart of the steps worth considering. The question is how to measure the achievement of the objective. An acceptable measure is the extent to which the courses of action are followed as the situation demands them. The library guide is probably insufficient for this purpose; better is a question-and-answer session in which students are required to answer test questions satisfactorily.

Knowledge of, or ability to find out about, the proper location in the library of specific items is closely linked to the previous objective and to some extent precedes it. Here too, the library guide is probably unable to give an acceptable measure of achievement of the objective.

The fifth objective - general impressions of material and staff - is not easy to measure, but it is necessary as an operational objective. The structure, quantity and quality of the literature are extensive and, to ensure its optimum and satisfactory use, librarians need to be effective (and seen to be so) as intermediaries between library material and users. Too often they are seen as being primarily interested in books rather than in users. Such librarians need to be balanced by others who are seen as "people", ie who are receptive to requests for assistance.

A final point related to the last objective is the interest, or lack of interest, in inanimate books.

A tour of a library can be compared with a tour of a large retail store or super-market. Many people, when drifting around shelves and looking at numerous inanimate objects, are likely to be bored unless their interest is captured; and that can be difficult if students do not accept the importance of libraries.

Thus the objectives can be to develop the following responses from students:

- I must come back and learn my way around this place.

- It is important to me to be able to use the library extensively.

- I know how to approach the problem of using the library.

- I know how to get assistance in using the library.

1.2 METHODS

The methods used by libraries for introducing the library to new students are varied and are likely to be determined by particular beliefs and circumstances. The variables are as follows:

- timing;
- duration;
- nature of groups;
- size of groups;
- use of guides;
- co-ordination of library tours;
- use of prepared presentations (or stock discussions);
- content - primary publications,
           - bibliographical tools,
           - library services,
           - procedures for use,
           - exercises.

1.2.1 Timing

The earliest time for consideration is when the students arrive at the university or college, and the latest time is after the students are firmly established on their courses.

It is the practice of a number of universities to spend a day or so establishing new students. The introduction to the library can be embodied in this procedure but it is unlikely that any substantial objectives can be established other than the distribution of readers' tickets and the location of the library. In short, if the university has a few days' orientation and the objective of the introduction to the library is very basic, it is acceptable. If the objectives are any more extensive, including loan procedures, stock division, and so on, they are being introduced too early.

Once the students are firmly established, an introduction to the library can be readily understood but it may be too late if the demands of the course have already necessitated the use of the library.

The change from a novel situation to a familiar pattern of study can take place in a few days. It would seem that the ideal time for library introduction is during this brief period of time, probably during the latter half of the first week and the early part of the second week at the university. The optimum time will depend on the detailed objectives of the introductory tour of the library and on local factors such as size and structure of the university.

There are practical difficulties in keeping to the optimum time for the tour; the number of students involved, conflict with other activities, and the ability of library staff to repeat material and yet capture interest. For a small organization with low student intake, there may be little difficulty, particularly as the library is likely to be on the small side. Conflict with other activities can usually be avoided by one means or another. The effects of repetitive and intense demand on library staff can be countered by using prepared material, including films and tape/slide presentations.

Finally, there are often good reasons for placing greater emphasis on other factors than timing. The main need, as suggested above, is to avoid being too early or too late for particular groups of students. In practice, the timing can be established in discussion with course tutors and departmental representatives.

1.2.2 Duration

The suggested duration is up to one hour. It can be reduced by incorporating the introduction in some other activity. If more than an hour is required, it is highly desirable to re-examine the library structure as it is likely to be too complex.

1.2.3 Nature of Groups

If the introduction to the library is in an orientation programme or is by invitation, mixed-subject groups are likely. Otherwise the groups can be conveniently organized by subject courses and the emphasis can be on subject-related material.

The arguments for mixed groups are that the literature provides extensive coverage and that interaction between students is good. At this level of instruction, the arguments are very slim.

1.2.4 Size of Groups

The ideal group size is one student, but this is normally impossible. A single student will participate in the library procedure and will relate the structure and organization of the stock to his needs and interests. Capturing his interest will be easy and the duration will probably be minimal because familiar material and material without immediate interest can be ignored.

At the other extreme, the complete student intake can be taken together. This is efficient in staff time, especially if prepared presentations using film or similar material are used. The problem is to achieve a reasonable objective for so mixed a group.

It is believed that somewhere between the two extremes a range of reasonable group sizes exists which permits reasonable objectives to be achieved with a reasonable expenditure of time and effort. The group sizes used will need to be based on experience and on the objectives chosen. If there is no earlier experience, it is suggested that group sizes should be limited to eight students if one intends them to participate and show interest. A small university of 3,000 students with a centralized library can cope with about five groups at a time - perhaps seven when co-ordinated and led by experienced guides. For library staff inexperienced in this work the group size should be less than eight.

These are fairly specific figures based on our own experience and belief. The arguments for using these are:

(a) Group sizes of eight students to one guide permit something of a seminar 'on location'. A seminar permits participation of each student and is more likely to be interesting and to achieve the objective of introduction to the library. The meaning of the word 'seminar', of sowing the seed of an idea, is suited to the objective of introducing the library as a place of interest and value, worth following up.

The factual material such as duration of loans, fines, etc., should be dealt with in a library guide, but there may be a need in some students' minds to appreciate the reasoning behind certain aspects, e.g. a heavy fine. This reasoning can easily be discussed in a seminar approach.

(b) Timetabled one-hour sessions around the first two or three weeks of term have been implied. Timetabling indicates an interest in subject departments and is likely to ensure better attendance.

(c) Groups based on courses give a compromise between handling the complete intake and dealing with a single student. In addition, one can achieve reasonable flexibility in timing. Discussion with subject departments will indicate when the first serious use of the library is expected.

1.2.5 Use of Guides

The obvious method of choosing guides is to assign one member of the library staff to each group. This method has proved satisfactory in practice. The main alternative is to assign staff to particular features, usually related to their normal duties. Thus, the cataloguers describe the catalogue, and periodicals staff the periodicals provision. The result is likely to be more formal than individually led groups; but it may permit a larger number of groups and a larger throughput.

Some librarians use subject specialization rather than functional specialization. Each subject specialist can advise his colleagues on particular features in the subject literature. By comparison, in the functional specialization, the specialist in information retrieval may be needed to advise on content.

Supplementary guides may be available - library-school students, or subject students at higher levels. If they are used, thought is required on how to convey the objectives, even though some students will have very realistic ideas about their needs and about problems of using libraries.

If there is a guide for each group, there is likely to be a limit to the number of hours he can lead seminars during one day. It may be as little as two or three hours.

1.2.6 Co-ordination of Library Tours

If a 'seminar on location' approach is used and the number of groups at any one time is limited, some co-ordination will be needed, especially if, for the practical reasons, a greater number of groups or other restrictions are imposed.

Co-ordination can pre-determine the route for each group and the time spent at each significant feature. If Guides are assigned to particular features, or prepared materials such as slides or displays are used, they can assist the co-ordination.

In the 'seminar on location' approach co-ordination is simply a matter of getting the right people in the right place at the right time, explaining what is expected of everybody, and assigning the groups.

1.2.7 Prepared Presentations

Prepared presentations are in some respects essential, in order to avoid giving unnecessary facts orally - which is inefficient, undesirable and expensive. Sometimes prepared presentations can be too extensive and thus lose the advantages of personal contact and being on location.

Essential prepared presentations include a printed library guide containing facts, and notices. If prepared presentations mean avoiding personal contact with library staff and visiting the library building itself, their achievement is relatively small.

Attention can be given to supplementary tours of the library, using any film or tape/slide presentations which are available. Care should be taken to ensure up-dating of the content.

Appendix 2

Handouts

1. BIBLIOGRAPHICAL TOOLS USED TO LOCATE INFORMATION

The tools can be locally produced to cover the collections of an individual library, or centrally published and generally available. The following notes are a simplification of a complex situation, there being for example considerable overlap of the various guides. There are subject guides to the literature but their usefulness varies considerably.

Group 1 Catalogues, Bibliographies and Book-lists

Essentially guides to books but some include lists of periodical articles, reports, etc.

(a) <u>Local</u>   - Library Catalogue

- used for checking the availability of a book in stock,
- used for immediately required subject matter,
- also useful for locating reference books and subject guides to the literature.

<u>Note</u>: Usually exists in two sections: one by name of author and one by subject or classification number (for the latter case an index is needed).

(b) <u>Published</u> - general bibliographies - e.g. <u>British Books in Print</u> and <u>Subject Guide to Books in Print</u>

- printed book catalogues - e.g. <u>Library of Congress - Books: Subject</u>

- Government Publications

- subject bibliographies - list of books etc. on a particular topic.

Group 2 Abstracting and Indexing Services

Essentially guides to periodical articles, but some specialize in other forms of literature such as news items, and many include several forms of literature.

(a) <u>Local</u>   - found in some special libraries and information services (e.g. <u>Particle Science and Technology Information Service</u>)

(b) <u>Published</u> - several thousand

- it is useful to be familiar with those in the areas relevant to one's particular subject interests.

Guides to abstracting services exist.

Group 3 Reference Books

Essentially distillation of significant data and factual information gathered from many sources (including literature, organizations, surveys and experiments). There are different types of reference books and these can be general (including geographically orientated) or subject-based. Thus, there are usually at least two alternative sections of the reference collection to be consulted. The various types of reference books include encyclopedias, dictionaries (including glossaries), handbooks, data compilations, statistical compilations, directories of individuals or organizations, trade directories, buyers' guides, timetables, telephone directories, maps, gazetteers.

Guides to reference books exist but tend to be slightly out of date and to increase searching time. The result is that many people prefer to browse in the appropriate sections of the reference collection.

2. MAJOR ABSTRACTING AND INDEXING JOURNALS

(a) Science and Technology

Applied Science and Technology Index
Biological Abstracts
British Technology Index
Chemical Abstracts
Dissertation Abstracts International, B: Sciences and Engineering
Electrical and Electronic Abstracts (Science Abstracts, series B)
Engineering Index
Government Reports Announcements (issued by U.S. Department of Commerce)
Index Medicus
International Aerospace Abstracts
Nuclear Science Abstracts
Physics Abstracts (Science Abstracts, series A)
Science Citation Index
Scientific and Technical Aerospace Reports (STAR)

(b) Social Science and Humanities

Anbar Abstracts
British Humanities Index
Business Periodicals Index
Dissertation Abstracts International, A: Humanities and Social Sciences
International Bibliography of the Social Sciences - Economics
International Bibliography of the Social Sciences - Political Science
International Bibliography of the Social Sciences - Anthropology
International Bibliography of the Social Sciences - Sociology
Keesing's Contemporary Archives
Readers Guide to Periodical Literature
Psychological Abstracts
Public Affairs Information Service Bulletin (PAIS)
Social Science and Humanities Index
Sociological Abstracts

B. LITERATURE SEARCH TECHNIQUES

1. STATE-OF-THE-ART LITERATURE SEARCH

This basic information is often required to build up knowledge and can be of the type:

"Recent information on the use of holography as a communication system"
or "Production of protein from methane".

The need can be for an essay, a feasibility study, a research or development project or 'trouble-shooting'.

Main difficulty is in selecting the right words (keywords) for searching the bibliographical tools.

Step 1 - Encode for Retrieval

(a) Write down the information required.
(b) Check the meaning of each keyword - dictionaries and encyclopedias.
(c) Note the synonyms of each keyword - dictionaries and encyclopedias.

Step 2 - Plan a Search Strategy - based on the search variables e.g. potential value of information; available time and resources.

What to search -

Reference Books	- known useful reference works
	- use subject index to the library's reference collection
	- browse appropriate classification sections of reference collection.
Books	- Library catalogues
	- Subject Guide to Books in Print plus the Library of Congress Catalogue - Books: Subject.
	- Subject bibliographies.
Periodical Articles	- Select appropriate major abstracting services.
	- Select appropriate specialized abstracting services.
Reports and Other Sources	- These have their own specific indexes, e.g. Scientific and Technical Aerospace Reports (S.T.A.R.) but the abstracting services now tend to cover this material.

Step 3 - Carry out the Search

Search most likely recent sources first:

Major abstracting services	- work retrospectively from most recent volume
	- note possible reference in full, and read or scan later
Specialized abstracts or indexes	- as above.

Fix limits to search - why information is needed is the important factor?
                     - good reviews and/or source paper will help.

Alternative and Additional Sources of Information

- Non-literature sources of information - organizations, consultants, observations, information services.
- Other forms of recorded information - e.g. newspapers, theses, patents.
- Alternative 'searching tools' - computerized information services, Science Citation Index.

2. DATA AND FACTUAL INFORMATION

This information may be required for decision-making or practical work and is often of the type: <u>What</u> is ...? <u>Who</u> is ...? <u>Where</u> is ...? <u>When</u> is ...? <u>How</u> is ...? <u>Why</u> is...? **Main difficulty is in selecting the appropriate reference books.**

Step 1 - Use Reference Collection

Select reference books according to the precise nature of information required (e.g. Who is ...? **indicates a reference** book such as <u>Who's Who</u>).

(a) **Search known works considered appropriate e.g.** <u>C.R.C. Handbook of Chemistry and Physics</u> **and Kempe's** <u>Engineers Yearbook</u>.

(b) Browse appropriate types of reference book in the class numbers of the subject, related subjects and works of a more general nature.

N.B. In some subjects good monographs on specialized topics exist in the main bookstock.

Step 2 - Use Abstracting Services (if appropriate)

Fix search limits and select appropriate abstracting services, (e.g. <u>Chemical Abstracts</u> is good for physical and chemical data).

Step 3 - Consult Information Service

Consult information service for your organization or, if necessary, approach directly those bodies and associations who are 'authoritative' in the subject field for the required information. Select the organizations from appropriate lists of organizations.

Step 4 - Determine Information by Observation (Experiment)

At this stage, information appears not to be available from oral or recorded sources. Data and facts can also be determined experimentally (a) when quicker, provided accuracy and reliability can be satisfactory; (b) when information is peculiar to local circumstances e.g. the height of a particular tree; the hardness of a particular piece of metal; (c) as experience for individual, or for testing equipment.

When considering published information, it is important to give attention to reliability, accuracy and evaluation, bearing in mind the following features:

(a) quoted sources of information
(b) stated limits of confidence and applicability
(c) comparative (evaluated) values.

Check the values - are they credible? Use more than one source for greater confidence.

3. SOURCES FOR SPECIALIZED INFORMATION NEEDS

<u>Product Information</u> - **Buyers' guides and appropriate** sections of subject-orientated reference works.
     - Technical indexes and similar services.
     - Market surveys in periodicals.
     - Organizations, e.g. design centres and some research institutes.

<u>Standards Information</u> - National Standards are often indexed by number and subject in published yearbooks.

<u>Materials</u> - Use reference books e.g. C. Mantell, <u>Engineering Materials Handbook</u>; Y.S. Touloukian (ed.), <u>Thermophysical Properties of High Temperature Solid Materials</u>.

<u>Metal Specifications</u> - **Use reference books e.g. R.B. Ross,** <u>Metallic Materials</u>, N.E. Woldman, <u>Engineering Alloys</u>, **or** <u>Stahlschlüssel</u>. **Failing this,** consult an appropriate research institute, if there is one.

<u>Patent Information</u> - Consult patent agent or information officer.

<u>News Items</u> - Use sources such as Keesing's <u>Contemporary Archives</u>, Index to the <u>Times</u>, <u>Research Index</u> (useful for commercial news).

<u>Theses</u> - Use sources such as <u>Aslib Index to Theses</u>, <u>Dissertation Abstracts</u>.

Appendix 3

Examples of extended lectures

A. PRODUCT INFORMATION

The purpose of this section is to outline a specimen lecture, with some background notes. As the product-information scene changes completely from country to country, any source examples given are from the United Kingdom and no attempt has been made at broader coverage. Students need first to be told the outline:

1. Definitions
2. User needs
3. Information sources.

Representative examples of the sources available within a particular country should be given.

1. DEFINITIONS

A wide variety of methods are used for disseminating product information. At the 'craft' end of the scale the only publicity may be visual display; at the other end various costly methods are used to advertise their products in the press. The products of groups, or industrial firms, may also be advertised in the press: if small, perhaps only in local media; but if large, in the national press and also on radio or television.

Some products of industry are complete equipments like an aeroplane or an ocean liner; others are components, or materials, or tools, which may be used to build saleable products or are in themselves saleable. Of these, some are ready for domestic use. The domestic or 'consumer market' rapidly changes and is subject to local variations in the brand names of products available, due to a whole range of factors not least of which is the cost of transport in distribution. It is not usual, therefore, to attempt to provide information on domestic consumer products, even on refrigerators, for example. Instead, the field is left to competitive advertising, although there are national organizations which publish comparative studies of consumer products. Also, some trade journals are information services in themselves; for example, The Grocer is a British periodical which includes current prices and ranges of available products.

Those components, materials, tools or instruments which are not ready or saleable are required by industrial firms, which wish to have the relevant data, such as properties, prices, or producers. When products are common, such as steel to a certain specification, data on properties may be found in reference handbooks or in standard specifications. Unfortunately, no-one appears yet to have produced an international concordance or comparative index to the various standard specifications that exist. Materials which start out as proprietary brands such as 'Lucite' (Du Pont, USA brand of polymethyl methacrylate) or its UK equivalent 'Perspex' may have their properties included under trade or chemical names in reference books containing materials data; but the manufacturers' names may not be included.

Industrial firms often advertise in the press and publish brochures or handouts or data sheets for distribution to potential customers. Such material is known as 'trade literature'.

Product Information

We are now in a position to define the nature of the product information which appears to be in most demand, at least in developed countries, by potential industrial or research and development customers. It includes information on product properties; or prices; or comparative data on several products; or manufacturers, agents or suppliers of known products. The forms in which companies publish information about their products may be sales brochures, technical data sheets or books, catalogues or lists (with or without prices), applications reports in which potential use situations are detailed, advertisements in the technical press or 'sales write-ups' in periodicals. Some of these media merely list products and some give information about them. There is considerable duplication in developed countries - often expensive duplication - in the forms of publicity or advertising used.

Users of Product Information

Users of the kind of information defined above are mostly purchasing officers of firms, professional persons (such as doctors or architects), research and development scientists or engineers (particularly design engineers), and management or technical consultants. Almost all firms are heavy users of the product information of others, as well as originators of information on their own products.

2. USER NEEDS

It follows that the kind of information needed varies, even for the same user in different situations; so does the amount of detail required. Students on the course can be asked to consider the following ways in which a requirement for company or 'trade literature' may arise:

(a) Search to ascertain whether a suitable special component, device or material is available on the market.
(b) Search for comparative data on similar items from different manufacturers in order to assist selection of the best for a given application.
(c) Search for any manufacturer or supplier of common components, devices or materials, subject to price, delivery time and other availability factors.
(d) Search for a specific, half-remembered reference to a new item which, if it could in fact be utilized, could assist the re-design of one's own existing product or contribute to an entirely new product.

Of these, (c) is normally easy to achieve, whether through office files, purchasing-department files, libraries or even the personal knowledge of a colleague. In (d) the phrase 'new item' is used as a reminder that current awareness of new products is at least as important as retrospective search for established products.

In practice (a), (b) and (d) are difficult questions which really call for comprehensive and very well indexed sources. Such sources apparently do not exist in the developed countries, though several governments make attempts to provide a central source in one form or another.

3. INFORMATION SOURCES

The sources available within a particular country, whether relating to its own products or to those of other countries, are bound to vary considerably. It is recommended that whatever sources are available should be described and their advantages and disadvantages discussed. With postgraduate groups, the students can actively participate in discussion, based on evaluation of available sources, according to the following criteria for an ideal product-information service:

(a) Comprehensiveness of manufacturer coverage.
(b) Comprehensiveness of item coverage within declared subject area.
(c) Availability of detailed subject (product) index.
(d) Up-to-dateness.
(e) Standardized methods of specification.
(f) Inclusion of sufficient data for the enquiry, e.g. prices.
(g) Minimum search delay.
(h) Availability of copies to take away (loan or retention).
(i) Reasonable cost.
(j) 'Current awareness' as well as 'retrospective search' capability.
(k) Data retrieval (i.e. specific facts) as well as reference retrieval (i.e. names of firms with relevant catalogues).

The aim of comprehensiveness is, of course, to reduce the number of separate sources consulted, and hence the time of search, to a minimum. A considerable amount of, for example, a design engineer's time is spent seeking product information, sometimes fruitlessly and never with any certainty of having conducted a full and reliable search. Human resources are not only the most important in all senses, but also tend to be expensive.

In Figure A.3.1 the kinds of sources available in the United Kingdom are set out as an example. For the less experienced, such sources can be described in a course session and questions or discussion invited. For the more mature and experienced students, such as postgraduates, specimens of sources can be discussed in relation to the criteria for an ideal product information service.

Figure A.3.1 shows that one may consult either suppliers, whose catalogues may cover the products of many firms, or manufacturers through their literature or their salesmen or both. As Figure A.3.1 should demonstrate, it is information about the manufacturers themselves that may prove difficult to locate.

PROPRIETARY PRODUCTS & MATERIALS

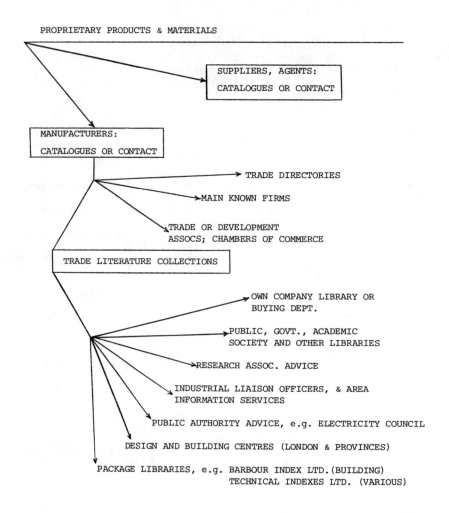

SUPPLIERS, AGENTS:
CATALOGUES OR CONTACT

MANUFACTURERS:
CATALOGUES OR CONTACT

TRADE DIRECTORIES

MAIN KNOWN FIRMS

TRADE OR DEVELOPMENT
ASSOCS; CHAMBERS OF COMMERCE

TRADE LITERATURE COLLECTIONS

OWN COMPANY LIBRARY OR
BUYING DEPT.

PUBLIC, GOVT., ACADEMIC
SOCIETY AND OTHER LIBRARIES

RESEARCH ASSOC. ADVICE

INDUSTRIAL LIAISON OFFICERS, & AREA
INFORMATION SERVICES

PUBLIC AUTHORITY ADVICE, e.g. ELECTRICITY COUNCIL

DESIGN AND BUILDING CENTRES (LONDON & PROVINCES)

PACKAGE LIBRARIES, e.g. BARBOUR INDEX LTD.(BUILDING)
TECHNICAL INDEXES LTD. (VARIOUS)

Figure A.3.1 : U.K. Sources of Proprietary Product Information

Guides to guides should be mentioned where appropriate, for example <u>Current British Directories</u> or the <u>Directory of British Associations</u>.

The difficulties with directories arise because of variations in availability of directory information in a given country or a particular subject field. The variations can affect the degree of subject breakdown, the comprehensiveness, the extent to which the information is still current, and the inclusion or omission of suppliers or agents, as well as actual manufacturers. Users may find a list of about fifty manufacturers under a broad subject heading and be faced with the unenviable task of consulting them all. Many specific trade or professional group-orientated directories, such as the <u>Engineer Buyers Guide</u> also include indexes of trade names, although some more general trade-name indexes exist. Of the general trade directories by country, the members of the <u>Kompass Register</u> series are the more comprehensive and have fairly detailed subject breakdowns.

The most detailed subject approaches which exist in published form are the indexes to 'package-library' services, such as the microfilm system of Technical Indexes Limited (UK) covering certain broad fields with specific subject indexing, or the Barbour Index Limited package-library system using a classification derived from the Swedish 'SfB' construction industry scheme and the relevant portions of the Universal Decimal Classification.

Those building up collections of trade literature must face variations in shapes, sizes and updating policies. Few firms date their literature, and for that reason will only supply price lists if specially requested. Even in large libraries the range of subject interests is so wide, and the number of firms so great, that only a selection can be maintained, though in such libraries one should at least find that the selection fairly represents the industries of the locality or the subject fields with which the library is concerned. Industrial libraries have to collect trade literature in their fields but can rarely afford sufficient staff time to organize it properly for retrieval. The collection or file may well lack, therefore, a detailed subject index. File organization may be by broad subject heading or class number, or simply by manufacturer's name. The last is the most common expedient, in which directories or package-library system indexes are used for the subject approach to the names.

Periodicals, even some of high technical level, may include advertisements, or articles contributed on behalf of manufacturers, or news items digested from trade literature sent to the editor. Certain periodicals are wholly or largely financed by advertising revenue (such as <u>Business Systems and Equipment</u>) and may be distributed free to potential customers, often including libraries: as such periodicals tend to cover broad subject areas they can be useful for current awareness, and it would seem unfortunate that their gaps, overlaps, variety of styles and lack of indexing are not corrected by co-ordination, which could of course be through voluntary co-operation between the publishers.

Of the requirements for product information stated earlier, (b) (Comparative Data) is probably the most pressing need. Technical books sometimes include lists of relevant manufacturers or even describe the products. A few other books go further and consist entirely of descriptions of components on the market, being therefore useful for comparative studies, although they start to become out-of-date as soon as they go to press.[1] Critical or evaluative comments are highly valued by the user but are not to be found in print, largely for legal reasons, except the studies of consumer products mentioned earlier. Advice is, however, sometimes available from research institutes, indeed, such advice is stated to be one of the aims of 'SIRAID', a service set up by the British Scientific Instruments Research Association. Public undertakings may also give advice, particularly on the application of their own products (electricity, for example).

When all directories fail, one may refer to trade associations, library collections or area information services, design or building centres, or even local chambers of commerce. However, as already noted, there are considerable variations in sources available within countries, and students should be encouraged to become familiar with the available sources which are important to his country, whether relevant to domestic products or imports. Practical work should be included, with test questions and preferably with supervision of the search exercises.

---

[1] For example, <u>British Miniature Electronic Components Data 1965-66</u>. G.W.A. Dummer and J.M. Robertson (eds.), London, Pergamon Press, 1965

In conclusion, the overriding limitations of any product-information source or service should be made quite clear to the student. Trade literature, or advice from experts, can only take an enquirer up to the point where it is necessary to make contact with a firm for full and up-to-date details, including prices, and possibly for the provision of specimens or demonstrations. The criterion regarding comprehensiveness, however desirable in order to minimize search time, is (or appears to be) impossible to meet completely. For example, small businesses tend not to be represented in the package libraries; the lack of coverage is strongly related to the amount they spend - or do not spend - on publicity and also to their tendency to work directly with an established market. Yet small firms are very important to the economy and to the social and industrial health of a country. Since many of them operate within fairly limited geographical areas, they can often be traced through classified telephone directories.

However complex the situation, product information searches do frequently narrow the field of potentially useful products for a particular purpose, though largely after considerable expenditure of time by professional or research and development staff. It is known that many searches prove unsuccessful or unreliable, or are abandoned through frustration, and the result can be financial loss due to marketing of items which are less than adequate or even already marketed in better form by another firm. Paradoxically, many firms spend considerable sums on publicity media designed to provide potential customers with product information free of charge.

These paragraphs have been included because they appear to be suitable for stimulating discussion, particularly with more senior students. The impression with which students of all levels should be left is that product-information searches, even when limited to proprietary products, can involve the use not only of trade literature but also of reference collections, textbooks, periodicals and individual experts, whether on the spot or in other organizations.

B. SPECIFIC TYPES OF INFORMATION NEEDS - STATISTICS

What are Statistics?

The distinction between data and statistics must be clearly understood. Data, as dealt with in Section 7.1, are numerical expressions of constant physical properties. Statistics are numerical records of social and economic phenomena, and are concerned with variables. Any organization may compile statistics of its own operations; a university knows, for instance, how many students are enrolled (total figures analysed by courses), and the size of its budget (total income analysed by sources; total expenditure analysed by uses). Similar records are kept in any kind of organization - manufacturing firms, shops, government offices, football clubs - and are an indispensable part of management information. Each individual statistic is a factual statement, but taken on its own it means very little. Only when it is considered in relation to other figures does it become significant. Comparison may be with other internal statistics, e.g. last week's production figures may be considered in conjunction with past weekly production figures. However, the local manager needs more than internally produced statistics to know how his enterprise is performing in comparison with other enterprises, and to discover such things as the size of the market for his firm's products. He needs statistics about other firms in his industry and about the potential users of his products (e.g. the agricultural engineer making ploughs wants to know the number, size and regional distribution of crop-growing farms).

Statistical returns from individual organizations or people are normally regarded as confidential. Therefore most published statistical information is presented for groups of people or organizations. These groups are known as aggregates. Thus individual returns are used to construct a broad picture with respect to the topic concerned. There can be many different levels of aggregation, depending on circumstances, but an over-riding requirement is that it must not be possible to identify statistics relating to individual people or organizations from published statistics. In many countries this is a legal obligation. Aggregates typically occur at the level of sub-divisions of an industry group (agricultural, motor vehicle, electrical with engineering), or of a class of products (ploughs, harrows, drills within agricultural machinery).

Who Publishes Statistics?

In any country the government is responsible for producing most of the publicly available statistics. What varies is the way the activities of collecting and publishing statistics are organized. In some countries these activities are highly centralized and conducted by a department named as a Central Statistical Office or Bureau of Statistics. In other countries they are left to individual government departments. (Course tutors should ascertain which pattern applies in their country.) For the user of statistics, the centralized system may be preferable, since there is more likely to be a single point to which information requests can be directed. The decentralized system requires the user to relate the subject of interest to the departmental structure of government. Organizations other than government departments which frequently collect and issue statistics include trade organizations, banks, research institutes and state-controlled industries. Of great importance, particularly in many developing countries, are the statistics collected and published by international organizations, especially the United Nations and agencies. For countries in which the national statistical service is at an early stage of development the fullest published statistics may exist in tables published by an international organization.

Types of Statistical Publication

The principal economic, social and demographic statistics for a country are usually published in a statistical year-book. In some countries this may be accompanied by a quarterly, or even a monthly, bulletin in which more up-to-date figures are presented. Most countries publish foreign-trade figures (imports and exports, with a breakdown by commodities). Usually other government statistical publications include census reports, covering population and possibly agriculture and industry. Censuses are usually taken at regular intervals, but these range between annual (usual for censuses of industries) and one every ten years (normal for censuses of population).

The reason for the pre-eminent position of the government in the publication of statistics is that the government needs statistics covering all aspects of life to enable it to discharge its responsibilities to the people of the country. Having collected this information, the government normally publishes it so that it is generally available.

When seeking statistical information, it is a good idea to ask two questions:

1. Is there a government department which would need this information?
2. Could it be collected as a by-product of a regular administrative procedure?

If there is a positive answer to one or both of these questions, it is likely to be that the information is being collected, and hence may be available.

Not all published statistics occur in publications which are specifically devoted to statistics. Reports on the activities of an organization, e.g. a government department, frequently include statistical tables. Also, there may be more than one source for a particular statistic. Official government statistics which are primarily published in an official publication are frequently published also in the press (newspapers and special-interest periodicals, such as trade journals), where they may appear within the text of an article commenting on their significance, rather than as tables. But these sources are much more difficult to keep track of and to identify, so it is in practice better to concentrate on the more obvious sources. These will also tend to be more reliable and more up-to-date.

Guides to Statistics

How is it possible for the would-be user of statistics to find out whether statistics have been published, and if so where? There are a number of guides to sources covering developing countries.[1] Information on what statistics are available may also be obtained from national statistical organizations, government departments and government publishers.

Any of these guides will lead the user to a particular statistical publication, in which the specific information required may be found within a table. Statistical publications are not always provided with detailed subject indexes, and it may be necessary to

---

[1] In particular a series by Joan M. Harvey (published by CBD Research Ltd, Beckenham, Kent, UK) which includes volumes on the statistics of Africa (1970), America (1973), Europe (1972) and Asia and Australia (1974). This series describes the major international and national sources, and the official statistical organization in each country. The United Nations has produced some useful guides, including a Guide to basic statistics in countries of the ECAFE regions (2nd ed. 1969. UN Economic Commission for Asia and the Far East). The GATT International Trade Centre in Geneva has published guides particularly intended for use by developing countries, although they are now somewhat dated - the Compendium of Sources: basic commodity statistics and the Compendium of Sources: international trade statistics having both appeared in 1967.

refer to a contents list giving table titles to identify relevant tables. As yet there is no indexing service providing detailed coverage of statistical sources even in the major industrial nations, though research is taking place in the United Kingdom (at Loughborough University of Technology) which may lead to the setting up of a system for United Kingdom statistical tables using techniques which could be applied in other countries. Statistics users the world over must appreciate that to locate required statistics, or even to verify that they do not exist, is often a difficult and time-consuming exercise. It pays to achieve familiarity with a small number of the most general statistical compilations and those which are most important for a subject area of particular interest, from which in practice it is likely that a high proportion of statistical enquiries can be answered.

The Use of Statistics

When a source for statistics has been found there are several questions which should be considered so that no misunderstanding of the information conveyed by the table occurs. These are:

1. What do the terms used in the table mean? Explanatory notes and definitions are often printed with the table or elsewhere in the publication. If a statistical classification is used, e.g. the Standard International Trade Classification in foreign trade statistics, reference to the classification schedules may help to resolve problems of definition and coverage.
2. What exactly do the figures given represent? Look for the statement of units of measurement. Note the area of coverage of the table and the date coverage.
3. What sort of figures does the table contain? Are they actual numbers, estimated numbers derived from a sample survey, index numbers (if so, what is the base year?), percentages, etc.?
4. If a series of figures is required over a considerable time period, can these be found in a single source? If not, it will be necessary to check very carefully that what appears to be an unbroken series of figures extending through several issues of a publication is in fact so. There may have been changes in the definition of concepts, or in the method of counting, or in the area covered.

Although much statistical information is published it is often possible to obtain additional tabulations from the compiling authority. It is therefore important to note the compiling body of a table which is of interest, so that it can be approached if additional information or clarification is required.

Notes for Course Tutors

In covering this topic the following may be useful exercises:

1. Ask each student to discover, for a topic of direct concern to him, what are the major sources of published statistical information, both national and international. How easy is it to make international comparisons, and to find comparable statistics over a period of five years, ten years and twenty years?
2. Draw up a list of the main compiling bodies at the national level in the country, and get from them any handout material describing their services and publications.
3. Consider how easy it is to gain access to a reasonably comprehensive collection of statistical sources? If there is no publicly available collection, can pressure be brought to bear on library authorities to do something?

For case studies of the uses of statistics it may be worth obtaining a UK publication Profit from Facts.[1] This could be used as a basis for constructing case studies suitable for local use.

---

[1] Obtainable from the Press and Information Service, Central Statistical Office, London SW1P 3 AQ.

The chapter on statistics and market research sources by Frank Cochrane in Manual of business library practice, edited by Malcolm J. Campbell (London: Clive Bingley, 1975) is a reliable source for further information on many topics touched on in this chapter, and for coverage of other aspects of statistical publications and their uses which it has not been possible to mention here.

Appendix 4

Examples of completed practical sheets

QUICK REFERENCE QUESTIONS

Name ....

1. Trace a book or introductory text on business ethics.

   GARRETT, T.M.   From:
   Business ethics   Subject Guide to Books in Print 1975
   Prentice-Hall, 1966   p.498

2. Trace additional details of a book entitled "Agricultural Marketing Boards: Their Establishment and Operation"

   ABBOTT, J.C.& CREUPELANDT, H.C.   From:
   (Food and Agriculture   Books in Print 1975, vol.3: Titles A-J
   Organization Marketing Guides: p.54
   No.5), Unipub. 1969

3. Give the number of cattle in Mexico, Sudan and West Malaysia.

   <u>1973</u>

   Mexico   26,548   From:
   Sudan   15,200*   U.N. Statistical Yearbook
   West Malaysia   326*   table 33, p.117-119

   (Figures in thousands of head)

   *Food and Agriculture Organization Estimates

4. Give the amounts of farm effluents that can be expected from dairy cattle, beef cattle and adult pigs.

   Dairy cattle   14 litres of liquid/day
                             51 litres of solids/day (includes straw)

   Beef cattle   7 litres of liquid/day
                             35 litres of solids/day (includes straw)

   Adult pigs   2.5 litres of liquid/day
                            15.0 litres of solids/day (includes straw)

                             From:
                             Kempe's Engineers Yearbook
                             p.1319

LITERATURE SEARCHING

Name ....

BRIEF: To study implementation of a new policy of extending the range of agricultural products. Any potential crops may be suggested but there is a known export market for paprika peppers.

1. Locate an introductory text which includes details on growing paprika peppers.

    ROYAL HORTICULTURAL SOCIETY
    Dictionary of Gardening, vol.   p.387

2. Locate any suitable material giving up-to-date developments (growing techniques, markets, use) and filling any gaps in the introductory text.

    Articles indexed in Biological Abstracts under generic names, e.g.
        CEPONIS, M.J. & BUTTERFIELD, J.E.
        Market losses in Florida cucumbers and bell peppers in
        Metropolitan New York.
        Plant Dis. Rep; vol.58, no.6, 1974, pp558-560.

        Retail losses in Capsicum frutescens var. grossum in seven
        New York retail stores totalled 9.2 per cent in three seasons
        1970-73. Wastage in consumer samples held for three days at
        39°F totalled 1.4 per cent. Decay, mechanical injury, chilling
        and desiccation were the major causes of losses.

3. Is the crop grown in your country?

    What is the botanical name for paprika pepper?
    Are there several varieties available? If so, which would you recommend?
    What crop yields would you hope for?

        a. Yes, introduced experimentally into St.Lucia in 1966
        b. Capsicum frutescens (= annuum)
        c. See Dictionary of Gardening and trade catalogues of seed suppliers
        d. 200 lbs/acre

4. Examine a potential country for marketing.

    What are its imports?
    Give the name of a possible importer and shipping line.
    What are the storage conditions?
    Does that country have relevant standards, laws?

        a. United States of America - incentive to experiment came from U.S. company.
        b. Detailed values not located in time and resources that were available.
        c. Baltimore Spice Co.
           For shipping lines see West Indies & Caribbean Year Book.
        d. 45 to 50°F and 90-95 per cent relative humidity can be waxed to reduce
           chafing in transit.
        e. As at January 1975
           - Customs invoice,
           - 3 copies of Bill of Lading.

5. Are there alternative crops which can be suggested?

    Nutmegs, Granada grows nutmegs in the wetter areas and exports all the
    products.

6. What organizations could advise and provide further information?

    Food and Agriculture Organization (FAO)

7. Are there any recent technical developments - irrigation, sowing, harvesting, drying, etc., which can be used to advantage?

    CANTLIFFE, D.J. & GOODWIN, P.
    Red color enhancement of pepper fruits by multiple applications of ethephon.
    J. Am. Soc. Hortic. Sci., vol.100, no.2, 1975, pp157-161

    (Numerous articles listed in Biological Abstracts)

Examples of Completed Practical Sheets

LITERATURE SEARCHING

NAME .... COURSE Mechanical Engineering (undergraduate)

	QUESTION	ANSWER
1.	Trace a text on fatigue of metals.	AUTHOR: FROST, N.E. & others TITLE: Metal fatigue PUBLISHER: Oxford University Press DATE OF PUBLICATION: 1975 SOURCE OF INFORMATION(& page):Subject Guide Books in Print 1975, p.2414
2.	Provide the bibliographical details of a book entitled: Engineering: an introduction to a creative profession	AUTHOR: BEAKLEY, G.C. & LEACH, H.W. PUBLISHER: MacMillan DATE OF PUBLICATION: 2nd ed.1972 SOURCE OF INFORMATION(& page):Books in Print, 1975, p.914
3.	What is the melting point of bismuth?	$271.3^{\circ}C$ SOURCE OF INFORMATION(& page): C.R.C. Handbook of Chemistry and Physics, 56th edition, p.8-76
4.	What is the detonating velocity of T.N.T. with a density of 1.64g/cc to be used for explosive forming of metal?	6940 m/s SOURCE OF INFORMATION(& page): PARRISH. Mechanical Engineer's Reference Book, p.9-3
5.	Trace an organization concerned with the galvanising industry in France.	Syndicat National des Fabricants d'Articles Galvanises et Etames. SOURCE OF INFORMATION(& page): Directory of European Associations, Part 1. p.88
6.	Give details of a recent article on burr-free drilling in soft materials.	TITLE OF ARTICLE: Burr-free drilling in soft materials AUTHOR: Anon. PERIODICAL TITLE: Manuf. Eng. & Mgt. VOL.NO:69 ISSUE NO:- DATE:Sept.1972 PAGES:24 SOURCE OF INFORMATION(& page): Applied Science & Technology Index 1973, p.381
7.	Give details of a recent article on the grinding of corundum.	TITLE OF ARTICLE: Grinding of corundum AUTHOR: SHAPIRO, M.K. PERIODICAL TITLE: Sov. J. Opt. Technol. VOL.NO:39 ISSUE NO:2 DATE:Feb 1972 PAGES:98-99 SOURCE OF INFORMATION(& page): Eng.Index 1973, p.802
8.	Give the name and address of a manufacturer of multigroove pulleys.	Aluminium Pulleys Ltd., Leafield Road, Shipton-under-Wychwood, Oxford, OX7 6EA SOURCE OF INFORMATION(& page):Technical Indexes Ltd., Engineering Components and Materials Index. April-Nov 1976, p. 145
9.	Give the number and date of the British Standard Specification for metric gauge blocks.	B.S. 4311:1968 SOURCE OF INFORMATION(& page): B.S. Yearbook
10.	What was the aluminium production of Canada in 1973?	930.2×10^3 metric tons of primary aluminium SOURCE OF INFORMATION(& page): U.N. Statistical Yearbook 1974, p.311

Examples of Completed Practical Sheets

LITERATURE SEARCHING

NAME .... COURSE Engineering Design (postgraduate)

Please give source of information with answer.

1. Find full details (author, publisher, date) of a book entitled: "Plastics and Synthetic Rubbers".
    GAIT, A.J. & HAWCOCK, E.G.   Published by Pergamon Press in 1970
    SOURCE: British Books in Print 1975, p.3273
2. Give the names of three abstracting journals which contain references in motor vehicle lighting systems.
    MIRA Automobile Abstracts
    Applied Science and Technology Index
    British Technology Index
    SOURCE: Library Subject Guide to Periodical Holdings, various pages.
3. Find the British Government publication reporting the findings of the enquiry into the accident at Markham Colliery.
    GREAT BRITAIN, Department of Energy
    Accident at Markham Colliery, Derbyshire, Report on the cause ...
    by J.W. Calder. (Cmnd. 5557)   Published by H.M.S.O. in 1974
    SOURCE: Government Publications 1974, p.1260
4. Who makes jam in Rotterdam?
    Co-op Nederland
    Rotterdam
    SOURCE: Kompass. Register of Netherlands Industry and Commerce, p.791
5. What organisation or individual is knowledgeable about tides?
    Institute of Coastal Oceanography and Tides
    Bidston Observatory
    Birkenhead
    Cheshire L43 7RA
    SOURCE:Directory of British Associations, p.156

LITERATURE SEARCHING

NAME ....COURSE Engineering Design (Postgraduate)

PROJECT: DESIGN A LARGE LAWN MOWER FOR GREENS AND GOLF COURSES

Manufacturers specification demands: (i) That the mower be hydraulically powered and self-propelled, but as light as possible. (ii) That the mower cuts with a traditional cylinder action to a width of 20" or more.

1. The starting point is the possible adaptation of a gang mower. Is there a book available which would provide a brief description of a gang mower - with an illustration if possible?
    CULPIN, C.   Farm machinery, 8th edition   Publ. Crosby Lockwood, 1969 pp306-323

2. Trace a directory listing manufacturers of gang mowers, so that one may be purchased as the basis for a prototype model.
    Farm and Garden Equipment Guide 1976   p130.

3. Which materials are best suited for use in the construction of a lawn mower? A periodical article may provide the answer.
    A number of associated periodical articles listed in A.S.T.I., B.T.I., and E.I. e.g. Aluminum wins over steel, magnesium, plastics in lawn mower. ANON. Mod. Metals, vol. 27, Sept., 1971, p44.

4. Trace a specification for agricultural mower parts which could be incorporated into the design of the new model.
    B.S.1562

5. Are there any fairly recent figures available on the value of the lawn mower production industry in the United Kingdom?
    See GREAT BRITAIN, Department of Industry, Business Statistics Office Report on the Census of Production PA 339.9 (Annual Statistics)

6. Large flail mowers are a possible source of competition.
    Is there a market survey on these, listing models available with prices, advantages, etc.
    The Green Book 1976, Section 25

Appendix 5

Select list of further reading

Two companion publications deal with the organization and evaluation of training courses and should be used for more detail on these aspects.

1. Unesco,"UNISIST Guidelines for the organization of training courses, workshops and seminars in scientific and technical information and documentation." Prepared by Pauline Atherton. Paris. 1975. 88p. (Doc.SC/75/WS/29)

2. Unesco, "UNISIST Guidelines for the evaluation of training courses, workshops and seminars in scientific and technical information and documentation". Prepared by F.W. Lancaster. Paris. 1975. 63p. (Doc.SC/75/WS/44)

Two significant works which can be read and consulted in conjunction with this Guide are:

3. LUBANS, J., Educating the library user. New York, Bowker. 1974. 435p.

A series of articles by numerous contributors gives a state-of-the-art review of the subject. It is useful background reading and contains some useful detail which is particularly valuable in the preparation of objectives and programmes. Although based mainly on U.S. experience, it offers a substantial range of coverage and viewpoints. The book has three parts, which indicate its use in preparation:

Part 1: Rationale for educating library users
Part 2: Faculty involvement in library-use instruction
Part 3: Implementation and evaluation of library-use instruction programmes.

Finally there is a selected bibliography.

4. PARKER, C.C.; Turley, R.V., Information sources in science and technology. London, Butterworth. 1975. 223p.

> Whereas Lubans is useful in overall planning, this volume provides invaluable information for lesson content and actual instruction. It can also be used satisfactorily for self-instruction. Teachers should find it a fruitful reference book not only for charts and handouts but also for following an established and proven programme of instruction. Details are given on a substantial number of search (finding) tools.

A list of the guides to the literature follows naturally on the practical approach of Parker and Turley. The following are commonly used guides:

5. CAREY, R.J.P., Finding and using technical information. London, Arnold. 1966. 153p.

6. CHANDLER, G., How to find out: a guide to sources of information for all; arranged by the Dewey decimal classification. 4th edition. Oxford, Pergamon Press. 1974. 194p.

7. CHENEY, F.N., Fundamental reference sources. Chicago, American Library Association. 1971. 318p.

8. GROGAN, D.J., Science and technology: an introduction to the literature. 2nd edition. London, Bingley. 1973. 254p.

9. HERNER, S., A brief guide to sources of scientific and technical information. Washington D.C., Information Resources Press. 1969. 102p.

10. HERNER, S.; MOODY, J. (comp.) Exhibits of sources of scientific and technical information. Washington D.C., Information Resources Press. 1971. 190p.

11. JENKINS, F.B., Science reference sources. 5th edition. Cambridge, Mass., M.I.T. Press. 1969. 231p.

12. LASWORTH, E.J., Reference sources in science and technology. Metuchen, N.J., Scarecrow Press. 1972. 305p.

13. MALINOWSKY, H.R., Science and engineering reference. Sources: a guide for students and librarians. Rochester, N.Y., Libraries Unlimited. 1967. 213p.

14. PARSONS, S.A.J., How to find out about engineering. Oxford, Pergamon Press. 1972. 285p.

There are in addition, guides to the literature and sources of information in particular subject fields. Many of these have been listed in Parker and Turley, pp. 42-45. Examples of those not included are:

15. Aslib; The Textile Institute, A Guide to sources of information in the textile industry. London. 1970.

16. PARKER, D.; CARABELLI, A.J., Guide for an agricultural library survey for developing countries. Metuchen, N.J., Scarecrow Press. 1970. 59,182p.

17. Schalit, M. (comp.), Guide to the literature of the sugar industry. Amsterdam, Elsevier. 1970. 184p.

It is impossible to list the bulk of useful references on various aspects of instruction. For further reading it is appropriate to start with the following and use them as indications for additional material.

18. International Conference on Training for Information Work. Rome, 15-19 November 1971. Proceedings. Ed. by G. Lubbock. Rome, The Italian National Information Institute; The Hague, International Federation for Documentation. 1972. 510p. (FID Publ. 486).

19. Use, Mis-Use and Non-Use of Academic Libraries. Proceedings of the New York Library Association - College and University Libraries Section Spring Conference held at Jefferson Community College, Watertown, May 1, 2, 1970.
New York College and University Libraries Section of the New York Library Association, c. 1970.
This is one of several publications in the early 1970's in which many of the basic issues are discussed. It has a good balance of viewpoints (a paper by a student is included) and an annotated and selected bibliography.

20. AVICENNE, P. (comp.), Bibliographical services throughout the world 1965-1969. Paris. 1972. 303p.

21. CAPSEY, S.R., Patents: an introduction for engineers and scientists. London, Newnes-Butterworth. 1973.

22. FINER, R. (comp.), A guide to selected computer-based information services. London, Aslib. 1972.

23. FINER, R.; BOWDEN, P.L. (comp.), A guide to selected British non-computer-based commercially-available information services. London, Aslib. 1975.

24. FOSKETT, A.C., A guide to personal indexes, using edge-notched, unic-term and peek-a-boo cards. 2nd edition. London, Bingley. 1971. 91p.

25. FOSKETT, A.C., The subject approach to information. 2nd edition. London, Bingley. 1971. 429p.

26. HARVEY, A.P., Directory of scientific directories. 2nd edition. Guernsey, Hodgson. 1972. 491p.

27. HOUGHTON, B., Technical information sources: a guide to patent specifications, standards and technical reports literature. 2nd edition. London, Bingley. 1972. 119p.

28. KASE, F.J., Foreign patents: a guide to official patent literature. Dobbs Ferry, N.Y., Oceana Publications; Leiden, Sijthoff. 1972. 358p.

29. LANCASTER, F.W.; FAYEN, E.G., Information retrieval on-line. Los Angeles, Melville. 1973. 597p.

30. TENNENT, R.M. (Ed.), Science data book. Edinburgh, Oliver & Boyd. 1971.

31. TEW, E.S. (Ed.), Yearbook of international organizations. 14th edition. Brussels, Union of International Associations. 1972.

32. Unesco, <u>World guide to science information and documentation services</u>. <u>Guide mondial des centres de documentation et d'information scientifiques</u>. Paris, UNESCO, 1965. 211p.

33. Unesco, <u>World guide to technical information and documentation services</u>. <u>Guide mondial des centres de documentation et d'information techniques</u>. 2nd edition. Paris. 1975. 515p.

34. Unesco, <u>Guide to national bibliographical information centres</u>. <u>Guide des centres nationaux d'information bibliographique</u>. Paris. 1970. 195p.

35. Unesco, <u>UNISIST: study report on the feasibility of a world science information system</u>. Paris. 1971. 161p.

36. WOOD, D.N.; HAMILTON, D.R.L., <u>The Information requirements of mechanical engineers; report of a recent survey</u>. London, Library Association. 1967. 35p.